DEVELOPMENT OF
A SMART TOURISM
DESTINATION

DEVELOPMENT OF A SMART TOURISM DESTINATION

Genka Rafailova
Zlatka Todorova-Hamdan
Hristina Filipova
University of Economics–Varna, Bulgaria

World Scientific

NEW JERSEY · LONDON · SINGAPORE · BEIJING · SHANGHAI · HONG KONG · TAIPEI · CHENNAI · TOKYO

Published by

World Scientific Publishing Co. Pte. Ltd.
5 Toh Tuck Link, Singapore 596224
USA office: 27 Warren Street, Suite 401-402, Hackensack, NJ 07601
UK office: 57 Shelton Street, Covent Garden, London WC2H 9HE

Library of Congress Cataloging-in-Publication Data
Names: Rafailova, Genka, author. | Todorova-Hamdan, Zlatka, author. | Filipova, Hristina, author.
Title: Development of a smart tourism destination: (using the example of Varna Municipality) /
　　Genka Rafailova, Zlatka Todorova-Hamdan, Hristina Filipova,
　　University of Economics-Varna, Bulgaria.
Description: New Jersey : World Scientific, [2025] | Includes bibliographical references and index.
Identifiers: LCCN 2024025328 | ISBN 9789811296086 (hardcover) |
　　ISBN 9789811296093 (ebook for institutions) | ISBN 9789811296109 (ebook for individuals)
Subjects: LCSH: Ecotourism--Bulgaria. | Smart cities--Bulgaria. | Sustainable tourism--Bulgaria.
Classification: LCC G155.B8 R35 2025 | DDC 910.68/4--dc23/eng/20241105
LC record available at https://lccn.loc.gov/2024025328

British Library Cataloguing-in-Publication Data
A catalogue record for this book is available from the British Library.

For any available supplementary material, please visit
https://www.worldscientific.com/worldscibooks/10.1142/13927#t=suppl

Desk Editors: Kannan Krishnan/Geysilla Jean/Analyn Alcala

Typeset by Stallion Press
Email: enquiries@stallionpress.com

Preface

The goal of this work is to study the evolution and implementation of the concept of a smart tourism destination to develop a human-centered model for a smart and sustainable destination 2030, which will be used to outline strategic directions of the development of the destination (as smart and sustainable).

In Chapter 1, the concept of a smart tourism destination (STD) is described in the context of its evolution from the idea of a "smart city." The definition of a smart tourism destination in three directions — fundamentals, processes, and results — is analyzed and determined. The interrelationship between sustainable development and STD development is also examined. Based on the theoretical justifications, an authors' model for a smart and sustainable tourism destination is elaborated, which rests on the existing approaches to the development of STD by linking its goals with the principles of sustainable development and the social aspect of their achievement.

Chapter 2 presents the methodology and toolkit for the research and evaluation of the smart and sustainable tourism destination (SSTD). The work presents the parameters of a survey conducted among three main groups of participants in the process of establishing a destination as sustainable and smart: experts, tourists, and citizens. The aim of this study is to determine the level of development of Varna, Bulgaria, as an SSTD. The opinions of respondents are united around the conclusion that the main problems are related to the achievements according to the indicators of mobility, accessibility, digitalization, and sustainable development.

Chapter 3 provides examples of successfully developed and established SSTDs in the cities of Valencia in Spain and Nice in France. The methodology for assessing the SSTD for the purpose of an in-depth analysis of the tourist destination, Varna, and the assessment of the cities of Burgas in Bulgaria, Thessaloniki in Greece, and Dubrovnik in Croatia as similar destinations is applied. Conclusions show the degree of development of Varna as an SSTD, which is defined as "a baseline with good potential," and the key achievements and opportunities, characteristics, and areas for improvement are outlined.

On the basis of the analysis, a proposal for the conceptualization of a strategy for the development of Varna as an SSTD has been derived. In the long term, the strategy is aimed at building and running a smart ecosystem for tourism, the creative sector, the ICT sector, and education and science as key and fundamental. This process is related to the development of innovative and sustainable maritime tourism within the blue economy and a creative tourism destination. This is attained through concrete activities and achievements in smart governance, sustainable development, and hard and soft smartness.

About the Authors

Genka Rafailova is Associate Professor (and former Director) at the College of Tourism in the University of Economics — Varna, Bulgaria, where she is also Head of the Centre for Cultural, Creative and Social Entrepreneurship. She holds a PhD in Tourism and has experience as an expert in European Union-funded international projects. Her fields of scientific interest include sustainable and smart tourism development, development of smart and creative destinations, creative industries and tourism, blue economy, and blue growth.

Zlatka Todorova-Hamdan is Chief Assistant Professor at the College of Tourism in the University of Economics — Varna, Bulgaria. She holds a PhD in Economics and Management (Tourism). She is a member of the Commission for Licensing of Tour Guides at the Ministry of Tourism, Bulgaria. Her fields of scientific interest include sustainable and smart tourism destinations, coastal tourism destinations, and cultural heritage and tourism.

Hristina Filipova is Chief Assistant Professor at the College of Tourism in the University of Economics — Varna, Bulgaria. She holds a PhD in Economics and Management (Tourism). She is a member of the Chamber of Tourism Varna and has experience as an expert in national and international research projects funded by ERASMUS + K2, etc. Her fields of scientific interest include consumer behavior, tourism marketing, and tangible and intangible cultural heritage.

Contents

Preface   v

About the Authors   vii

List of Figures   xi

List of Tables   xiii

List of Commonly Used Abbreviations   xvii

Introduction   xix

Chapter 1   Essence and Characteristics of a Smart
             Tourism Destination   1
             1.1. The Concept of a Smart Tourism Destination   1
             1.2. Models and Methodologies for Research
                  of an STD   23
             1.3. Sustainable Development of Tourism, Tourist
                  Destinations and Their Management   31
             1.4. Smart and Sustainable Tourism
                  Destination Model   42

Chapter 2   Research and Analysis of a Smart and Sustainable
             Tourism Destination (Municipality of Varna)   51
             2.1. Methodology of the Study   51
             2.2. Experts' Research: Analysis and Evaluation
                  of Results   65

2.3. Analysis and Evaluation of the Results
      of a Survey of Tourists' Opinions   73
2.4. Analysis and Evaluation of the Results
      of a Survey of Citizens' Opinions   82

Chapter 3   Guidelines for the Development of Varna
             as a Smart Tourism Destination 2030   93
             3.1. Analysis of Successfully Developed Smart
                  Tourism Destinations   93
             3.2. Assessment of Varna as an SSTD   104
             3.3. Comparative Characteristics of Varna, Burgas,
                  Thessaloniki and Dubrovnik as SSTDs   127
             3.4. Strategy for the Development of Varna
                  as an SSTD 2030   151

Conclusion   157

Appendix A: Models and Methodologies for STD Research   161

Appendix B: Indicators for the Assessment of an SSTD   167

*Appendix C: Comparison of Cities Based on SSTD
             Evaluation Indicators*   179

Bibliography   181

Index   193

List of Figures

Fig. 1.1.　The concept of a smart city　8

Fig. 1.2.　Smart tourism destination　21

Fig. 1.3.　Model for a smart and sustainable tourism destination　47

Fig. 2.1.　Methodology of the study　52

Fig. 2.2.　Methodology for assessing a SSTD　52

Fig. 2.3.　Summary of the assessment of the use of modern smart technologies in the business of the surveyed experts　66

Fig. 2.4.　Distribution of answers to the question "To what extent are QR Codes used in Varna for better and faster access to information about tourist attractions and places?"　69

Fig. 2.5.　Distribution of answers to the question "Are smart technologies used in Varna?"　70

Fig. 2.6.　Characteristics of the behavior of tourists in Varna: Frequency, duration and purpose of visit　74

Fig. 2.7.　Assessment of accessibility to sights in Varna for people with special needs　76

Fig. 2.8.　Assessment of the level of development of Varna as an STD based on the main indicators (tourists)　81

Fig. 2.9.   Distribution of citizens' responses related to
                electronic administrative services   85

Fig. 2.10.   Distribution of answers for the indicator
                "Cultural heritage and creativity" (citizens)   87

Fig. 3.1.   Comparative assessment of Varna, Burgas,
                Thessaloniki and Dubrovnik as SSTDs   150

Fig. 3.2.   Strategy for the development of Varna as an
                SSTD 2030   152

Fig. A.1.   System model of a smart tourism destination   161

Fig. A.2.   Smart tourism destination framework   162

Fig. A.3.   Structure of model of intelligent destination:
                Field and criteria   164

Fig. A.4.   Theoretical directions of transforming smartness
                into sustainability   165

Fig. A.5.   A sustainable smart tourist destination model   166

List of Tables

Table 1.1.   Data on urban population (1990–2020)   3

Table 1.2.   Definitions of a smart city (2000–2022)   4

Table 1.3.   Number of nights spent by tourists in short-stay
accommodation through online booking on shared
travel platforms, such as Airbnb, Booking.com,
Expedia Group and TripAdvisor (2018–2019)   10

Table 1.4.   Definitions of an STD (2015–2022)   12

Table 1.5.   Fundamental technologies for the development
of smart destinations and tourism   14

Table 1.6.   Principles for the sustainable development of a
tourist destination   36

Table 1.7.   Indicators for the sustainable development of a
tourist destination: Environmental sustainability   39

Table 1.8.   Indicators for the sustainable development of
a tourist destination: Economic and social
sustainability and effective destination management   41

Table 2.1.   Description of the survey among stakeholders
in Varna   62

Table 2.2.   Thematic parts of the questionnaire for experts
and their role in the survey   63

Table 2.3.   Thematic parts of the questionnaire for tourists
and their role in the survey   64

Table 2.4.   Thematic parts of the questionnaire for citizens
and their role in the survey   65

Table 2.5.   Summary assessment by indicators of partnership
between stakeholders in tourism   67

Table 2.6.   Summary assessment of the elements of smart
infrastructure and smart mobility in Varna   68

Table 2.7.   Summary assessment of the management of Varna
as a tourist destination   71

Table 2.8.   Mode and mean of the elements of the "accessibility"
criterion for Varna (tourists)   75

Table 2.9.   Mode and mean value of the elements of the
indicator "Mobility" for tourists in Varna   77

Table 2.10.   Mode and mean value of the elements of the
indicator "digitalization" for Varna (tourists)   77

Table 2.11.   Mode and mean of the elements of the indicator
"Sustainable Development" for Varna (tourists)   78

Table 2.12.   Mode and mean value of the elements of
"Cultural heritage and creativity" criterion for
destination Varna (tourists)   79

Table 2.13.   Ranking smart destination indicators by
importance (tourists)   80

Table 2.14.   Mode and mean of the elements of the
"accessibility" indicator for Varna (citizens)   83

Table 2.15.   Assessment of accessibility of tourist sites for
different target groups   84

Table 2.16.   Mode and mean of the elements of the "mobility"
indicator for Varna (citizens)   84

Table 2.17.   Mode and mean of the elements of the indicator
"digitalization" for Varna (citizens)   85

Table 2.18. Mode and mean value of the elements of the
indicator "sustainable development" for destination
Varna (citizens)   86

Table 2.19. Ranking by relevance of smart destination
indicators (citizens)   87

Table 2.20. Results of the comparative analysis of means for
tourists and citizens (fourteen SSTD criteria)   90

Table 2.21. Similarity of the mean values for citizens and
tourists for the different criteria of SSTD   91

Table 3.1.   Number of hotels and rooms in the city of
Nice (2020)   103

Table 3.2.   Number of beds and accommodation establishments
in Varna Municipality (2019–2022)   112

Table 3.3.   Number of beds per 100 people of the population
in Varna (2019–2022)   126

Table 3.4.   General comparative information about the
destinations Varna, Burgas, Thessaloniki and
Dubrovnik (2022–2023)   129

Table 3.5.   Quality of life in the destinations of Burgas,
Varna, Thessaloniki and Dubrovnik (2023)   131

Table A.1.   STD study areas and categories   163

Table B.1.   Baseline indicators for the assessment of SSTDs   167

Table B.2.   Indicators for assessment of SSTDs   171

Table B.3.   Indicators for assessment of SSTD for exclusivity   175

Table C.1.   Comparison of Varna, Burgas, Thessaloniki and
Dubrovnik based on SSTD evaluation indicators   179

List of Commonly Used Abbreviations

AI	Artificial Intelligence
AR	Augmented Reality
BAS	Bulgarian Academy of Sciences
DMO	Destination Management Organization
GDP	Gross Domestic Product
IoT	Internet of Things
MBT	Mechanical-Biological Treatment
MR	Mixed Reality
NGOs	Non-Governmental Organization
NSI	National Statistical Institute
PM_{10}	Particulate Matter 10 Micrometres or Less in Diameter
QR Code	Quick Response Code
R&D	Research and Development
RES	Renewable Energy Sources
RFID	Radio-Frequency Identification
RIEW	Regional Inspection of Environment and Water
RSO	Regional Statistical Office
SMEs	Small and Medium-sized Enterprises
SSTD	Smart and Sustainable Tourism Destination
STD	Smart Tourism Destination
TIC	Tourist Information Centre
VR	Virtual Reality
WWTP	Wastewater Treatment Plant

Introduction

This work presents a summary of the theory and good practices, research results, and models and strategies for the development of a smart tourism destination (STD) 2030 using the example of Varna, Bulgaria.

The relevance and significance of the problems and research related to the development of an STD are determined by the following factors and trends:

(1) the contemporary complex and dynamic socioeconomic and political environment for the development of tourism and tourist destinations;
(2) the existing need and requirements for sustainable development, including tourism and destinations;
(3) the dynamic development and application of information and communication technologies (ICT) in all social and economic areas;
(4) intense competition between destinations based on the experience offered;
(5) the values and lifestyle of modern generations — orientation to experiences;
(6) the desires, needs and interests of modern tourists and changes in tourist demand.

The current environment in which tourism develops is marked by complex political conflicts and real and potential threats of pandemics and epidemics. This changes travel conditions primarily by imposing restrictions or requirements that travelers must comply with. The economic environment is characterized by a number of unfavorable trends, such as rising inflation and logistical difficulties with fuel and labor shortages. This, on the one hand, creates fear and uncertainty that prevents people from traveling. On the other hand, it requires the provision of real-time information, the guarantees of protection and security, and the offer of an even more valuable experience for tourists.

Today, more than ever, the management of tourism destinations and tourism businesses is required to apply the principles of sustainable development. Some factors determining this need include deepening urban problems, overtourism, seasonality of tourist trips, pollution caused by the tourism sector and conflicts with the local community and the ongoing destruction of biodiversity. The solutions are related to the development and implementation of models and policies oriented toward the Global Goals for Sustainable Development. At the same time, scientific advances and innovations allow the deployment of technologies ("green," "harmless," and "safe") that reduce or prevent negative environmental impacts, contribute to the development of a circular economy and achieve carbon neutrality and a better quality of life. Their use becomes a factor for gaining competitive advantages, preserving the attractiveness of the destination and compliance with international agreements and regulatory requirements. The dynamics in ICT and their applications in all spheres of public life increase the dependence of competitiveness on digitalization and change the lifestyle. Accordingly, the successful development of tourist destinations is determined primarily by the character of the experiences offered and not just by the location and the available natural resources and tangible cultural heritage. The use of "green" and digital technologies is a factor for the competitiveness of the destination thanks to the achievement of an economic effect, preserving the natural environment and anthropogenic material values and creating a positive image.

Digitalization, globalization and an increased standard of life are changing the lifestyles of contemporary generations. Their preferences and interests are related to interactive and hybrid experiences, healthy lifestyles and the use of "green" technologies and products. They have increased requirements for information availability and the opportunity to control the activities they carry out. This, of course, significantly changes the tourist demand. Tourists want the offered tourism products to enable active participation, learning combined with fun and being informed easily, quickly and constantly, but above all security and safety during trips.

The concept of a smart tourist destination, which evolved from the concept of a "smart city" and based on modern ICT, is developed and applied primarily to contribute to solving the urban problems caused by the growth of the tourism sector, such as pollution, overcrowding and heavy traffic (Rafailova *et al.*, 2022). But, it is oriented to achieve effective partnerships between stakeholders, creating a higher quality experience for tourists and raising the standard of living of the local community. A smart tourist destination is defined by Gretzel *et al.* (2015) as a tourism system that has advantages generated by the use of smart technologies related to the creation, management and delivery of smart services/experiences for tourists and is characterized by intensive information sharing and joint development of products and values. One of the essential aspects of this concept is its interrelationship with sustainable development. According to one of the basic definitions of an STD of SEGITTUR (Segittur, 2015), its main feature is the creation of conditions for the sustainability of a destination. However, numerous studies related to the development and application of this concept have found poor achievements in terms of sustainable development, mainly related to higher energy efficiency and waste management. The positive results relate primarily to the application of modern ICT for the purpose of increasing tourist satisfaction and the popularity of the destination, but without ensuring a better quality of life for residents in the destination. In this sense, it is necessary for the development of a tourist destination as smart to be defined by a strategy in which the objectives and principles of sustainable development are clearly integrated and are socially oriented, i.e., to take the interests and needs of the local community and tourists into account at the same time.

The subject of this study is the development of a tourist destination as smart and sustainable and the achievement of competitiveness, quality of life for the local community and the experiences of tourists.

The object of this study is a tourism destination: Varna.

The goal of this work is to study the evolution and application of the concept of an STD to develop a human-centered model for a smart and sustainable destination 2030, which will be used to outline strategic directions of destination development (as smart and sustainable).

In order to achieve this objective, the following research tasks are set:

(1) a review and analysis on the theory and practice behind the development of an STD and sustainable development in tourism;

(2) bringing out the characteristics of a smart and sustainable tourism destination (SSTD);

(3) development of a methodology for assessing the level of the development of an SSTD;

(4) application of the developed model and methodology to the Municipality of Varna and the implementation of related studies;

(5) determining the level and directions of development of the city of Varna as an SSTD.

The methods and approaches of this study used include the method of analysis and synthesis; the method of deduction and induction; the methods of empirical research, including in-depth and structured interviews with experts, questionnaires and analysis of statistical data, and research of secondary data, such as policies, documents and good practices.

The main limitations of this study are as follows:

(1) The subject of this study requires a study of a wide range of theory and literature that are closely related to it — the nature and role of digital technologies, the legal consequences of their application and the contemporary aspects of urban planning and development, which, however, cannot be thoroughly covered and examined within this work.

(2) Studies of good practices are limited by the availability of sufficient data and information, primarily for destinations similar to the city of Varna, and are implemented for a limited number of destinations.

(3) The study of tourists' opinions and ratings is limited by the possibilities of contacting them, especially those who use Airbnb accommodations, Booking.com rooms and suites, and the like.

(4) Due to the lack of sufficient information, the indicator "business competitiveness" has not been considered in Chapter 3.

Chapter 1 presents a theoretical overview and analysis of the characteristics of "smart" cities, smart tourism and smart tourist destinations, sustainable development of tourism and tourist destinations, aspects of urban planning and the management of a tourist destination and application of modern digital technologies in tourism, including for the goals of sustainable development and development of a smart destination and city. On this basis, a model for an SSTD is derived, which is human-centered.

Chapter 2 describes the methodology of the research and the nature and results of the conducted studies on the opinions and assessments of experts, tourists and residents in Varna.

Chapter 3 sets out the conclusions of the studies, including good practices, and presents a socially oriented strategy for the smart and sustainable development of a tourism destination 2030.

This monograph was prepared in the implementation of the research project NPI-52/18.05.2021 on the topic: Development of a Smart Tourism Destination 2030 (Using the Example of Varna Municipality). The project is funded at the expense of a subsidy from the state budget and has a period of implementation from 18.05.2021 to 31.12.2022. The study was carried out with the participation of lecturers, PhD students and students from the University of Economics, Varna.

The contributions of the authors to the development of this work are as follows:

(1) Associate Professor Genka Ivanova Rafailova: Chapter 1, Sections 1.2, 1.3 and 1.4, and Chapter 3, Sections 3.2 and 3.4.
(2) Chief Assistant Zlatka Ivanova Todorova-Hamdan: Chapter 1, Section 1.1, and Chapter 3, Sections 3.1 and 3.3.
(3) Chief Assistant Dr. Hristina Plamenova Filipova: Chapter 2, Introduction and Conclusion.

The authors would like to express their gratitude to the Municipality of Varna, the Tourism Directorate and the Varna Tourist Centre, as well as to representatives of the tourism business in the municipality and Antonio Hadzhikolev for his contribution to the technical implementation of the studies and layout of the monograph.

Chapter 1

Essence and Characteristics of a Smart Tourism Destination

1.1. The Concept of a Smart Tourism Destination

The concept of a smart tourism destination (STD) has evolved from the concept of a "smart city" with the dynamic development of information and communication technologies (ICT) and the growing need to find complex solutions, including those related to the development of tourism, based on multiple and diverse data. The terms "intelligent" and "smart" in this work correspond to "smart"[1] (e.g., smart destination, smart town and smart tourism), which is used in the scientific literature in English. But, they are primarily associated with the theoretical and widely accepted interpretation of the concepts of "smart" and "smartness," expressing the connection with the environment through the Internet, applying an innovative approach to solving problems and responding to a variety of situations based on digital technologies, knowledge from different fields and real-time information (Gretzel, 2015; Vargas-Sánchez, 2016).

The concept of a "smart city" was established at the beginning of the twenty-first century due to the growing problems of urbanization and the

---

[1] Cambridge Dictionary, https://dictionary.cambridge.org/dictionary/english/smartness — "smart" and "smartness": It means and refers to the quality of intelligence, for quick thinking, the ability to behave intelligently in complex situations, and the translation of the word "smart" in Bulgarian is *умен, интелигентен*.

need for a change in the model of urban management, oriented toward the goals and principles of sustainable development and at the same time increasing the attractiveness for investment and qualified human resources. The population of the world's cities (see Table 1.1) increased to 56% between 1990 and 2020. Although their average annual growth is gradually decreasing, it is almost double the average annual growth of the entire population of the earth. Territorially, the trend is in the direction of an increase in the number of cities, at an average of 11% per year. Forecasts for 2030 and 2050 outline similar trends. These are some of the reasons why global urban planning and development address a number of challenges, especially in terms of resource efficiency and territory, air pollution and social services.

An essential role in changing the model of urban governance and development is played by the progress and accelerated application of ICT in all spheres of public life. In the world (Statista, 2022), the global spending on information technology — devices, databases, business software and communication and information services (the largest share of investment) — reached $4.43 trillion in 2022, up to 14% from 2020, and is expected to grow by 5% to reach $4.67 trillion in 2023, continuing accelerated digitalization in various fields. As a result of these processes, the concept of the "smart city" was created, which in the period 2000–2022 was approved and changed (see Table 1.2). However, the overview of the different definitions shows that the basic idea of intelligent urban development is complementary and upgraded.

The initial focus of the concept of a smart city is the modernization of its infrastructure and its management (Hall, 2000), but above all to improve the service to the residents. The essence and role of a smart city are considered in two directions (Moura *et al.*, 2017): technocratic and socially oriented. The first group includes definitions and research with a focus on the application of ICT and digitalization of infrastructure, management and administrative services, economy and environmental impact (Su *et al.*, 2011; Lombardi *et al.*, 2012; Gue *et al.*, 2017; Peng *et al.*, 2017; Boot, 2018; Moura and Silva, 2019; Ringel *et al.*, 2021). The second group includes definitions and studies aimed at improving the lifestyle and wellbeing of local residents (Giffinger *et al.*, 2007; Thuzar, 2011; Lazaroiu and Roscia, 2012; Calderon *et al.*, 2018; Borruso and Balletto, 2022). The development and application of the concept of a smart city are carried out in relation to economic and environmental sustainability (BSI, 2014; Moura and Silva, 2019) and partly social sustainability.

Table 1.1.   Data on urban population (1990–2020).

	1990	1995	2000	2005	2010	2015	2020
City population (thousands)	2,290,228	2,575,505	2,868,308	3,215,906	3,594,868	3,981,498	4,378,994
Average annual rate of change of urban population		(1990–1995) 2.35	(1995–2000) 2.15	(2000–2005) 2.29	(2005–2010) 2.23	(2010–2015) 2.04	(2015–2020) 1.90
Share of population in cities	40%	42%	44%	49%	52%	54%	56%
Average annual population growth		(1990–1995) 1.58	(1995–2000) 1.38	(2000–2005) 1.30	(2005–2010) 1.26	(2010–2015) 1.23	(2015–2020) 1.11
Number of cities with a population of more than 300,000	991	1,115	1,291	1,446	1,599	1,774	1,934

Source: Developed by the authors according to UN data, Department of Economic and Social Activities, Population Division, www.population.org (last visited August 17, 2022).

Table 1.2. Definitions of a smart city (2000–2022).

Author (year)	Definition
Hall *et al.* (2000)	A city that monitors and integrates conditions of all its critical infrastructures, including roads, bridges, tunnels, railways, subways, airports, seaports, communications, water, power and even major buildings, can better optimize its resources, plan its preventive maintenance activities and monitor security aspects while maximizing services to its citizens.
Giffinger *et al.* (2007)	A city well performing in a forward-looking way in economy, people, governance, mobility, environment and living, built on the smart combination of endowments and activities of self-decisive, independent and aware citizens. The smart city generally refers to the search and identification of intelligent solutions which allow modern cities to enhance the quality of the services provided to citizens.
Harrison *et al.* (2010)	A city connecting the physical, IT, social and business infrastructures to leverage the collective intelligence of the city.
Su *et al.* (2011)	The product of the digital city combined with the Internet of Things.
Caragliu *et al.* (2011)	A city is smart when investments in human and social capital and traditional (transport) and modern (ICT) communication infrastructure fuel sustainable economic growth and a high quality of life, with a wise management of natural resources, through participatory governance.
Thuzar (2011)	Cities that have a high quality of life; those that pursue sustainable economic development through investments in human and social capital, and traditional and modern communications infrastructure (transport and information communication technology), and manage natural resources through participatory policies.
Lazaroiu and Roscia (2012)	A community of average technology size, interconnected and sustainable, comfortable, attractive, and secure.
Lombardi *et al.* (2012)	The application of information and communications technology (ICT) with its effects on human capital/education, social and relational capital, and environmental issues is often indicated by the notion of a smart city.

Cretu (2012)	Two main streams of research ideas: (1) smart cities should do everything related to governance and economy using new thinking paradigms, and (2) smart cities are all about networks of sensors, smart devices, real-time data and ICT integration in every aspect of human life.
Bakici *et al.* (2013)	A high-tech intensive and advanced city that connects people, information, and city elements using new technologies in order to create a sustainable, greener city, competitive and innovative commerce, and increased life quality.
Dameri (2013)	A smart city is a well-defined geographical area where high technology such as ICT, logistics, energy production, etc. cooperate to create benefits for citizens in terms of wellbeing, inclusion, participation, environmental quality and smart development. It is governed by a well-defined set of entities capable of defining the rules and policy for urban management and development. The most important entities in the definition of a smart city should be citizens; however, they are often overlooked. Implementing a smart city initiative means not only achieving technological success but also using technology to create societal value.
British Standards Institution (BSI) (2014)	The effective integration of physical, digital and human systems into the built environment to ensure a sustainable, prosperous and inclusive future for its citizens.
Marsal-Llacuna *et al.* (2015)	Smart-city initiatives try to improve urban performance by using data information and information technologies (IT) to provide more efficient services to citizens, to monitor and optimize existing infrastructure, to increase collaboration among different economic actors, and to encourage innovative business models in both the private and public sectors.
Peng *et al.* (2017)	A city using a set of advanced technologies, such as wireless sensors, smart meters, intelligent vehicles, smartphones, mobile networks or data storage technologies.
Guo *et al.* (2017)	Urban development based on the integration of many ICT solutions to manage the city's resources.
Dameri (2017)	The level of smartness of a city could be defined based on its core components: land (or territory), infrastructure, people and government.

(*Continued*)

Table 1.2. (*Continued*)

Author (year)	Definition
Booth (2018)	Smart cities will depend increasingly on data-intensive technologies and processes, making the cloud a critical requirement.
Calderon *et al.* (2018)	A city would be considered "smart" if it applies ICT-based solutions to problems in these six dimensions: governance, quality of life and essential services, transportation (mobility), economy, people, and environmental issues. Furthermore, a city must include initiatives and projects in its development plans to be considered as a city becoming smart.
Moura and Silva (2019)	A smart city can be defined as an urban area (encompassing possibly different areas and scales of the city — street, plaza, neighborhood, or, ultimately, an entire city) that uses electronic data collection sensors located in infrastructures, buildings, vehicles, institutions and devices (Internet of Things) to supply real-time information of the main cities' operating systems. In the setting of a smart city, city authorities can also promote participatory government to ultimately promote more sustainable urbanized development and a more competitive and attractive business and creative environment.
Yigitcanlar *et al.* (2019)	A city is smart when investments in traditional infrastructure, social development and modern (ICT) communication infrastructure fuel sustainable growth and a high quality of life, with the wise management of natural resources.
Ringel (2021)	A technology-intensive city that delivers "intelligent" energy and mobility solutions in cooperation with its citizens. We add the dimension of "sustainability," implying both the minimization of resource streams and environmental impacts as well as adaptation to a changing global climate.
Borruso and Balletto (2022)	The Smart City, at present, should be a city that tackles the needs of its citizens and city-users without focusing on their technological skills or devices. It should provide solutions through technological infrastructure and devices, such as smartphones and apps, now widely used, but the true change consists of putting the ICT-related procedure "under the bonnet" and embedded into the processes, and therefore not necessarily visible and detectable by the single citizen. The term therefore evolved in time from more purely digital aspects to wider ones.

Source: Based on Moura and Abreu e Silva (2019).

This concerns both the search for smart solutions to improve the environmental impact of decarbonization[2] (Ringel, 2021), pollution reduction (Bakici *et al.*, 2013) and the optimal use of resources (Caragliu *et al.*, 2011; Guo *et al.*, 2017; Yigitcanlar *et al.*, 2019) and so for economic prosperity (Thuzar, 2011; Caragliu, 2011), social integration (Dameri, 2013) and democratization of urban governance (Caragliu *et al.*, 2011; Moura and Silva, 2019; Mora *et al.*, 2019). The integration of digital technologies into urban infrastructure and utilities, community life and administrative services is seen as significant for a smart city in terms of networking, partnership and connectivity between stakeholders (Lazaroiu and Roscia, 2012; Bakici *et al.*, 2013; Marsal-Llacuna *et al.*, 2015).

The concept of a smart city is realized through the development of its key areas (Giffinger, 2007; Dameri, 2017; Calderon *et al.*, 2018; Galati, 2018): "smart economy" (competitiveness and sustainable growth), "smart governance" (open to broad participation), "smart mobility" (physical, transport and social), "smart environment" (conservation and environmentally friendly use of resources), "smart people" (knowledgeable, independent and informed people) and "intelligent lifestyle" (valuable). All of them are bound by the goal of improving the quality of life and the environment. In these key directions, the assessment of the average (between 100,000 and 500,000 inhabitants) and after 2015 of the larger (between 300,000 and 1,000,000 inhabitants) European cities as smart and their ranking (Centre of Regional Science, Vienna UT, 2007) begin.

A smart city is distinguished by the digitalization of infrastructure, economy, transport, education, health and other spheres of social life, smart governance and investments in human resources, leading to a higher quality of life and a sustainable environment (see Fig. 1.1). The application of digital technologies is oriented toward improving the service of city residents and their mobility, reducing and preventing negative impacts on the environment and maintaining effective connectivity between stakeholders. Urban governance is carried out through the use of large databases and the active participation of citizens and businesses. A good quality of life is based on both an attractive environment for work and leisure and quality education and social integration. In a smart city, the efficient and optimal use of resources is achieved.

---

[2]Decarbonization is a process of decreasing the carbon emissions, the main factor for climate change.

Pillars

Digitalization
infrastructure, economy, transport, administrative services, spheres of social life,

Intelligent governance
shared, large databases oriented towards sustainable development

Smart Investments
human resources, innovation,

Purposes

Quality of life

A stable environment

Results

Smart mobility

Quality environment for work and leisure

Stakeholder connectivity and social integration

Efficient use of resources

Ecological impact on the environment

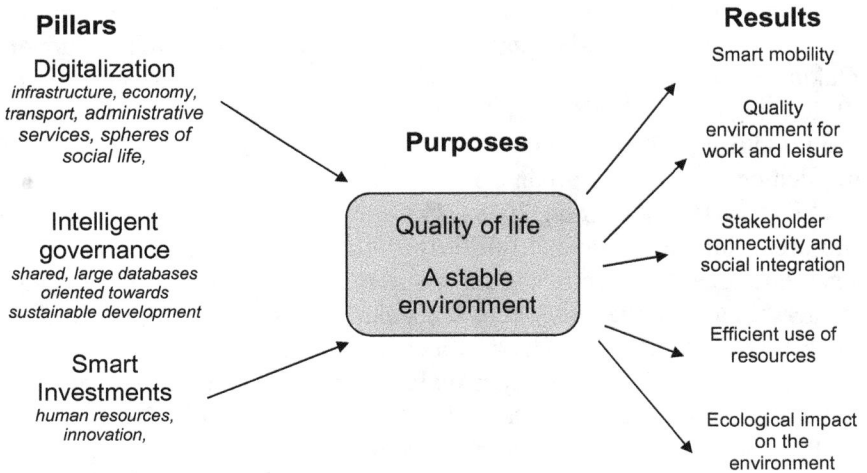

Fig. 1.1.   The concept of a smart city.

Source: Developed by the authors based on Moura and Abreu e Silva (2019) and Ringel (2021).

The orientation of the concept is in the direction of building and maintaining an ecosystem (Appio *et al.*, 2019; Borruso and Balletto, 2022), which allows a high degree of connectivity between the key stakeholders in the city, and the innovative development and implementation of smart solutions both to improve the quality of life and address today's economic, social and environmental challenges, including global climate change.

Logically, from the concept of a smart city comes smart destination and smart tourism. One of the main reasons is the need to solve similar problems related to the development of the tourism sector. Tourism has a strong impact on the natural and physical anthropogenic environment — it influences natural resources, factors and functions — and the human-made environment for the purposes of production, trade, recreation and leisure, provision of services, living, etc. (Inskeep, 1991; Cooper and Archer, 1994; Pearce, 1994; Coccossis and Nijkamp, 1995; Hunter and Green, 1995; Kaspar and Mihalic, 1996; Newsom *et al.*, 2002; Holden, 2001; WTO, 2001; EC, 1997). Numerous studies reveal a multitude of negative consequences and processes in tourism destinations, mostly caused by overtourism and the seasonality of tourist trips — exceeding the carrying capacity — natural and social, noise, heavy traffic, pollution (including architecturally), urbanization of natural areas, destruction of valuable resources, destruction of esthetic appearance and beautiful views, violation of ecological functions, etc. (Cooper *et al.*, 1993; Bieger, 2002; Bosselman *et al.*, 1999; Pearce, 1994; Hunter and Green, 1995; Newsom *et al.*, 2002; Vodenska, 2001; Holden, 2001; Tosun, 2001; WTO; EC). At the same time, business, a major consumer of natural and anthropogenic resources, on which it actually depends (Goodall and Stabler, 1997; Buchalis and Fletcher, 1998), together with the organizations responsible for the management of tourist destinations, is looking for opportunities and approaches to meet the tourism demand — dynamic and changing under the influence of digital technologies — and to build appropriate competitive advantages in the face of intense competition.

According to the WTO report on transport emissions (World Tourism Organization and International Transport Forum, 2019), in 2005, tourism's contribution to human-made carbon emissions worldwide was 5%, with a transport share of around 70%. The data and forecast for tourist trips are as follows: an increase from 770 million in 2005 to 1.2 billion in 2016 and reaching 1.8 billion in 2030 or a 45% increase compared to 2016. Domestic travel increased from four billion in 2005 to eight billion in 2016 and is projected to reach 15.6 billion in 2030, a 78% increase compared to 2016. At the same time, in 2016, passenger transport accounted for 64% of total transport emissions, and despite increasing fuel efficiency and green transport modes, the growing demand for passenger and freight transport will lead to higher levels of carbon dioxide emissions by around 21% in 2030. In 2030, tourism-related transport will contribute to 23% of the world's carbon emissions from transport and 5.3% of all human-induced emissions. Therefore, despite the positive role of tourism in economic development (10% of global GDP and 10% of employment), the sector will continue to have a significant impact on the environment and climate change.

One of the significant problems related to tourism in Europe is its seasonality, which contributes to overtourism in many destinations. According to Eurostat (2022) data, one-third of overnight stays and one in four trips are in July and August. Contributions to the concentration of tourists during the summer season have platforms for shared travel and accommodation, such as Airbnb (see Table 1.3).

The data show an increase in interest in bookings through online platforms offering short-stay accommodation, which has a positive impact on an increase in tourists to the destination. At the same time, this way of staying leads to a deepening of the problems of overtourism

Table 1.3. Number of nights spent by tourists in short-stay accommodation through online booking on shared travel platforms, such as Airbnb, Booking.com, Expedia Group and TripAdvisor (2018–2019).

	Number of nights spent by tourists in the EU 27	Number of nights spent by tourists in July and August in the EU 27	Number of tourists per night in the EU 27	Number of nights spent in other accommodation establishments in the EU 27
2018	441,915,000	159,655,000 (39%)	1,182,000	4,755,414,647
2019	511,939,000	182,682 (37%)	1,400,000	4,765,035,752

Source: Developed by the authors based on Eurostat data, https://ec.europa.eu/eurostat/statistics-explained/index.php?title=Short-stay_accommodation_offered_via_online_collaborative_economy_platforms (last visited August 20, 2022).

and contributes to exacerbating the negative impact of seasonality on tourist demand. According to the above data, one-third of the nights spent through shared travel platforms occur, as in the other places for accommodation, during the months of July and August, which is traditionally the high season in Europe. Online booking platforms give travelers not only more freedom and choice of accommodation before traveling but also more flexibility during the visit to the destination. Despite the insignificant share of nights spent compared to those in other places for accommodation, it is necessary to note that the concentration of tourists, with reservations on these platforms, is primarily in cities (43 cities in the EU 27 and EFTA with over one million nights), mainly capitals such as not only Paris, Lisbon, Rome, Madrid, Budapest, Prague, Vienna and Athens but also Barcelona, Porto, Milan and Nice.

A key and fundamental definition of STD was proposed back in 2015 by Lopez de Avila, which is used in the smart tourism project in Spain (SEGITTUR — a Spanish state-owned company for innovation and technology in tourism, 2015) and the subsequent research:

STD is an innovative tourism destination built on the basis of high-tech infrastructure ensuring the sustainable development of tourist places, accessible to everyone and in which the connection and integration of tourists in their environment is facilitated, the quality of the experience is increased and the quality of life of the local population is improved.

This outlines some of the most important features of STDs: the use of high (digital) technologies, including for the purpose of protecting the environment, creating a smart experience for visitors and a good/high standard of living for the local community. Despite the wide applicability and breadth of the definition, numerous concepts and theoretical formulations about STDs are subsequently derived, which enrich and adapt it to the new socioeconomic and technological realities.

The definition of an STD in the scientific literature, in studies with the aim of developing policies and strategies and in international initiatives and evaluation and ranking systems is carried out in three main directions (see Table 1.4). The first of these is the definition and study of the foundation of STDs: modern infrastructure (mainly for modern ICT, parking and transport and waste treatment), digital technologies and big databases (Gretzel *et al.*, 2015, 2016). The second direction presents the STD processes that characterize and distinguish it: resource efficiency (Buhalis and Amaranggana, 2013), reducing the negative impact on the environment (Gretzel *et al.*, 2015; González-Reverte, 2018), ensuring and enhancing mobility, accessibility and creativity (EC, 2022), urban governance through the involvement of multiple actors (Lamfus *et al.*, 2015), networking stakeholders (Buhalis and Amaranggana, 2014) and the co-creation of values (Boes *et al.*, 2015; Vargas-Sánchez, 2016). The third direction defines and explores the results of the development of STDs: creating a more qualitative, exciting and "smart" experience for tourists (Neuhofer and Buhalis, 2014; Boes *et al.*, 2015) and a higher and more valuable quality of life (Gretzel *et al.*, 2015; Vargas-Sánchez, 2016; Ivars-Baidal *et al.*, 2019; González-Reverte, 2019).

The foundation of STDs not just encompasses modern digital technologies but is also determined by their integration into the destination infrastructure (Gretzel *et al.*, 2015; Lamfus *et al.*, 2015). These are technologies that provide not only the constant connectivity of tourists and residents with an environment, between key actors in the tourism sector, economy and society, but also the opportunity to make and apply smart decisions at different levels and in different areas. For example, among ICT planning applications are computer simulations that help visualize future scenarios and develop forecasting techniques and geographic information systems that can capture, store, analyze and display large amounts of geographic data. Computerized systems of indicators that allow for the monitoring and control of the territorial planning of tourism are available. An additional contribution is made by mobile devices that provide

Table 1.4. Definitions of an STD (2015–2022).

Year	Author	Definition
2015	López de Ávila	STD is an innovative tourist destination, built on the basis of high-tech infrastructure, guaranteeing sustainable development of tourist places, accessible to everyone and with facilitated connection and integration of tourists in their environment. As a result, the quality of the tourist experience and the quality of life of the local population are increased.
2015	Gretzel *et al.*	Sum of all integrated efforts in the destination to apply smart technologies for environmental sustainability, increased mobility, inclusion and innovation.
2015	Buhalis and Amaranggana	Specific smart-city options that take advantage of smart technologies to provide added value to tourists.
2015	Boes *et al.*	Places that use available technological tools and techniques to enable supply and demand to jointly create value, satisfaction and experiences for tourists, profits and benefits for organizations and the destination.
2015	Lamfus *et al.*	A tourist destination is intelligent when it makes intensive use of the technological infrastructure provided by the smart city to: (1) enhance visitor experiences by personalizing and familiarizing both local and tourism services and products in the destination, and (2) by empowering destination managers, local institutions and tourism companies to make decisions and take action based on the data, collected, managed and processed using technological infrastructure within the destination.
2016	Gretzel, Zhong and Koo	A new smart tourism economy with new resources, new players and new models of exchange based on opportunities provided by smart technologies.

2016	Vargas-Sánchez	It is based on the extensive use of modern technologies to create an advanced digital space through an integrated network of management systems, platforms and in short all kinds of data. The goal is to improve the overall management of the destination.
2017	Li *et al.*	Individual tourist information system for information services in the context of global technologies.
2018	Jasrotia and Gangrotia	If a smart city uses information technology and innovation to improve tourism's six As, it eventually becomes a smart-tourist destination. In other words, smart-tourist destinations are the cities or places that use available technologies, tools, innovations and techniques to provide enjoyment and experiences for the tourist, and profit for organizations and destinations.
2022a, 2022b	European Commission	A smart-tourist destination is defined as a destination where different stakeholders, possibly under the coordination of a destination-management organization, facilitate access to tourism and hotel products, services, spaces and experiences through innovative ICT-based solutions, making tourism sustainable and accessible, and making full use of their cultural heritage and creativity.

Source: Developed by the authors based on Gretzel *et al.* (2015, 2016), Buhalis *et al.* (2014, 2016), Vargas-Sánchez (2016) and EC (2022).

access to the Internet and, respectively, information, reservations and purchases, etc.

They are considered according to their role in serving tourists and locals, creating a tourist product, managing the destination and tourism, interconnection with the environment, etc. Table 1.5 summarizes and presents the key technologies and systems for a smart destination and city. Considered in this aspect, a smart destination is the basis for the development of smart tourism, which has evolved from e-tourism along with the progress of ITC (Gretzel *et al.*, 2015), the digitalization of the entire value chain in tourism and the relationship between physical sites and digital infrastructure (Gajdošík, 2018; González-Reverte, 2019). Smart tourism is distinguished by experiences for tourists in which they actively participate and are offered according to their interests and needs and the time and place (Femenia-Serra and Neuhofer, 2018; González-Reverte, 2019)

Table 1.5.   Fundamental technologies for the development of smart destinations and tourism.

Technology	Nature and role	Scopes per app
Internet of Things (IoT)	Connection between objects via the Internet, between a person and the surrounding environment, with less interference from people	Collecting data, transmitting information according to context, location and specific needs, and maintaining connectivity between people and the environment
iBeacon	Protocol developed by Apple; the technology allows smartphones and tablets to perform various actions when near an iBeacon transmitter	iBeacon sensors for buying tickets for transport through the smartphone, receiving in an urban environment relevant information directly on tourists' mobile devices, personalized experience in museums, etc.
Cloud Services	Cloud for Iaas, PaaS, SaaS, IoTaaS, AIaaS	Cloud infrastructure services for access to information regardless of place and time, for tourists and businesses
RFID — Radio Frequency Identification	Technology for the transmission and processing of digital information (data)	Identify multiple objects at the same time, unlike barcodes where scanning objects happens one at a time

Table 1.5.   (*Continued*)

Technology	Nature and role	Scopes per app
QR codes	Encrypted information that is read by scanning from a smart device	Share tourist maps, locations, images, links and PDFs for easy Internet access in the hotel, in the museum to get information in several languages, etc.
Sensors	Reception of information from the surrounding environment, including by people and their mobile smart devices	Maintaining an urbanized environment, signaling crowding, trafficking, disasters and problems
Machine learning	A functional part of artificial intelligence that allows computers to make decisions without being pre-programmed for them	Opportunity for the tourism and transport sectors to obtain solutions to long-standing business problems related to customization, pricing, optimization of business processes and forecasting consumer behavior
Artificial Intelligence	Artificial intelligence	On-the-Process Automation, Chatbot platforms, virtual tour guide
AR	Augmented reality — a combination of digital information with a real object and environment	Visualization and presentation of cultural heritage
VR and MR	Visualization of real objects or the creation of such in the virtual space; mixed reality and metaverse	Branding and advertising of tourist destinations and businesses, interpretation and presentation of cultural heritage, and new entertainment

Source: Derived by the authors, revised and supplemented, from Todorova-Hamdan and Hadzhikolev (2021), based on Hsu and Tsaih (2018).

thanks to access to real-time information and collaboration between multiple partners, including local government.

The foundation of an STD is its "hard smartness"[3] (Vargas-Sánchez, 2018; Boes *et al.*, 2016). It covers an infrastructure with implemented

---

[3] The infrastructural and technical aspect of the smartness.

digital technologies for socioeconomic activity, utility, information and communication services, including the maintenance of big databases. At the same time, it should be taken into account that the development of a tourist destination is determined by six pillars: attractions, tourist super-structure, accessibility, offered tourist packages and services, activities and ancillary services (Buhalis, 1998; Boes *et al.*, 2016; Gajdošík, 2018), designated as the "Six As."[4]

Providing a high-tech environment does not guarantee the develop-ment of a destination as smart (Vargas-Sánchez, 2016; Perles-Ribes and Ivars-Baidal, 2018; Gajdošík, 2018). The effective application and use of digital technologies require knowledge and cooperation between stake-holders and appropriate management, including through public–private partnerships (Perles-Ribes and Ivars-Baidal, 2018). Otherwise, the mech-anism for the development of an STD is "soft smartness"[5] (Boes *et al.*, 2016; Gajdošík, 2018), defined by the competencies for the creation and implementation of innovations and by a managerial approach driven by the capabilities of big databases (Vargas-Sánchez *et al.*, 2018). In this aspect, a number of studies are aimed at defining the processes that take place in an STD. They cover policies and activities for the smart develop-ment of the tourism sector, related to ecology, mobility, education, innova-tion and the management of cultural and historical heritage. A significant difference between e-tourism and smart tourism (González-Reverte, 2019) lies in not only the creation of joint value in the context of the loca-tion and the desires of visitors but also offering it in accordance with environmental protection and resource efficiency. These processes and activities are supported and implemented by the smart management of the destination, represented by local authorities and the organizations respon-sible for its management. It is a governance that has a clear and publicly accessible approach to decision-making in cooperation with stakeholders and through the use of big databases, with electronic administrative services involved in public–private partnerships (Rafailova, 2020; Buhalis and Amaranggana, 2014; Gretzel, 2018). Smart destination gov-ernance is an essential factor for value co-creation (Femenia-Serra and Neuhofer, 2018), which is the core of the intelligent experience

---

[4]Attractions, amenities, accessibility, activities, available packages, and ancillary services, as accessibility is treated as an opportunity for getting to the destination and its attractions, type of transport, quality and availability of infrastructure.

[5]The functional expression of the smartness.

(Buhalis *et al.*, 2014). This process is based on the concept of product creation through the partnership between the user and the manufacturer (Prahalad and Ramaswamy, 2004). A significant contribution to the evolution of this cooperation between the client and the creator includes modern technologies, social networks (Facebook, Twitter, Instagram, TikTok, YouTube) and communication processes of "content" (shared platforms, blogs, company mobile and navigation applications) that change the relationships between users, producers and intermediaries and turn them into partners (Tussyadiah and Fesenmaier, 2009; Buhalis, 2009; Neuhofer *et al.*, 2013; Buhalis *et al.*, 2023). Subsequently, their roles are more fully explained by the theory of the logic of service dominance (S-D logic) (Vargo and Lusch, 2008). According to the basic principle of this theory, values jointly created by multiple actors (Kryvinska *et al.*, 2013) always involve the beneficiary, and all other social and economic partners are contributors of resources in the form above all of knowledge and skills (Vargo and Lusch, 2008). In the theory of the S-D logic, co-creation is seen as a process of building a complete experience for tourists in which they are actively involved together with a tourism business, DMO and other actors in an economy and social life in the destination, as well as the local residents. It is a dynamic exchange and at the same time consumption of information and resources and services between business and public organizations, tourists and businesses and tourists and public organizations. This process is carried out through the constant connectivity of key actors and tourists, joining open platforms and providing access to databases (Buhalis, 2015). On the one hand, the digitalization and integration of ICT in all phases of the value chain — before, during and after the journey — are a prerequisite for this process. On the other hand, smart governance supports the co-creation of values by encouraging and building a network of stakeholders and information exchange between them, developing public–private partnerships and maintaining open databases. Local residents also participate in this process through information and knowledge and/or acting as partners in creating the experience by providing services, training and tours as part of their daily lifestyle.

An essential element of a smart destination is environmental sustainability (Perles-Ribes and Ivars-Baidal, 2018; EC, 2022), which is defined by the achievement of a more efficient use of resources (González-Reverte, 2019), increasing energy efficiency, reducing the climate change consequences of tourism development and preserving cultural heritage and local values. In the scientific literature and applied projects on smart tourism, mainly in urban environments, these processes are associated with transport management oriented toward "emission-neutral" or "green" transport and traffic (reduction of congestion and infrastructure load), control over crowds of people and visitors in certain places (González-Reverte, 2019), particularly valuable attractions, as well as the use of "green" energy technologies in public spaces and institutions (Ivars-Baidal *et al.*, 2021). The sustainability of the environment, an element that is transferred from the concept of the smart city (Lee *et al.*, 2020), is a priority for the development of a smart destination, but above all in the environmental aspect — counteracting pollution, creating green-belts and protecting natural resources — as well as in the social aspect, but mainly in terms of controlling overtourism and prolonging the season by attracting tourists from socially vulnerable groups and seniors aged above 65. Despite the undeniable importance of these processes, they have a short- or medium-term impact and are carried out for the purpose of the efficiency of destination marketing and the competitiveness of the tourism sector. According to González-Reverte (2019), a destination is considered smart if it is sustainable. In practice, however, this direct connectivity is not always achieved and sustainability is not set as a key priority and goal of strategies for the development of smart cities (Philipova, 2022).

The theoretical overview reveals two key orientations of smart urban development and tourism destinations, respectively: a smart experience for tourists and a high quality of life for locals (Boes *et al.*, 2015). The factors that determine the targeting of these results are as follows:

- the lifestyle and preferences of the current generation of tourists (González-Reverte, 2019), especially millennials[6] (Nieradka, 2016), who intensively use digital technologies and online services for entertainment, education, purchases and information and are looking for an opportunity to be in constant contact with the environment and people, including being informed in real time;
- new technological solutions and digital technologies that allow better-quality health and education services, creating online communities

---

[6]The millennial generation or generation *Z*. The use of mobile applications in the group of generation *Y*. Managing innovation and diversity in knowledge society through turbulent times. *Joint International Conference Timisoara*, Romania, p. 815.

and sharing data, opinions, experiences and user participation in the services offered (Prahalad and Ramaswamy, 2004);
• the humanization of urban policies.

A smart experience is a personalized experience for tourists according to their preferences and characteristics, based on modern ICTs that provide easy, fast and environmentally friendly access to services, attractions and real-time information, tailored to place and situation, and participation in value creation (Femenia-Serra and Neuhofer, 2019). This complex of impressions, sensations and emotions related to the destination and its attractions, entertainment, cultural heritage, tourist services and infrastructure is characterized by creativity, innovation and durability of impact, including after visiting the destination. In an STD, tourists easily perceive the information offered without language restrictions (Tsaih and Hsu, 2018) and have freedom of movement and more time for entertainment, recreation and individually preferred activities (Buhalis and Amaranggana, 2015). The goal of a smart experience is to provide more positive emotions and lead to a higher degree of satisfaction for both visitors and locals. Theoretically, the views and perceptions of the two groups — tourists and local populations — are taken into account, but little research has looked at the interrelationships between them. In the approaches and policies applied to the development of a smart destination, the goal of an intelligent, satisfying experience is not always set in relation to residents, but rather in parallel with it, thus planning to achieve a higher quality of life. Despite the importance of these results, it is necessary to take into account the role of the local population as a participant and at the same time a user of co-created value in tourism. At the same time, tourists should be considered "temporary residents" (Gajdošík, 2018).

The quality of life is determined by the comfort and esthetics of the urban environment, access to quality social services, health care and education and the provided opportunity for residents to participate in decision-making, directly affecting their present and future (Gretzel *et al.*, 2018).

In many urban tourism destinations in Europe (EC Study, 2020) — Amsterdam (iBeacon Mile), Gothenburg (Event Impact Calculator), Helsinki (real-time crowding heatmap) and London (Digital LITH, Smart London Plan), especially those characterized by the negative impacts of overtourism such as Venice (smart monitoring system), Florence

(Silfi Smart City Control Room), Costa del Sol (secure destinations dashboard), Seville (intelligent tourism destination) and others — smart systems (with sensors, iBeacons, machine learning, etc.) are implemented to monitor, control and regulate transport traffic and movement of people, especially in areas with valuable attractions, such as architectural, archaeological and other monuments. Thanks to these systems, higher environmental sustainability, better accessibility, preservation of cultural and historical heritage and higher satisfaction of tourists and locals are achieved. For the purposes of improving the interactive relationship with visitors and increasing their active participation in the creation of the experience, in a number of European destinations, e.g., Berlin (virtual experiences), Brasov (augmented reality application), Florence (Museo dell'Opera interactive visit), Lisbon (shops with history), Brussels (neighborhood walks) and Valencia (city past view) — smart solutions based on innovative technologies are used to present a story, navigation, acquaintance with the location and more.

Based on the theoretical overview and the derived foundation, processes and results of the development of STDs, the authors propose the following definition:

A smart tourist destination is a destination with modern infrastructure, integrated digital technologies in a socioeconomic environment and smart governance, which are the foundations of cooperation and connectivity between stakeholders, tourists and locals for environmental sustainability, leading to the creation of an intelligent tourist experience and an increase in the quality of life (see Fig. 1.2).

The key determinants of an STD or STD foundation cover the following:

(1) **Modern infrastructure:** This includes infrastructure for transport, utilities, business and living, equipped with sensors and allowing remote monitoring and control, providing easy and quick access when moving.

(2) **Digitalization of the socioeconomic environment:** This includes digitalization of provision of services in health care, the social sector, education and transport, implementation of digital technologies for communications, provision of access to the Internet, creation and maintenance of big databases.

(3) **Smart governance:** This involves "shared" governance with a wide range of stakeholders, based on digital technologies.

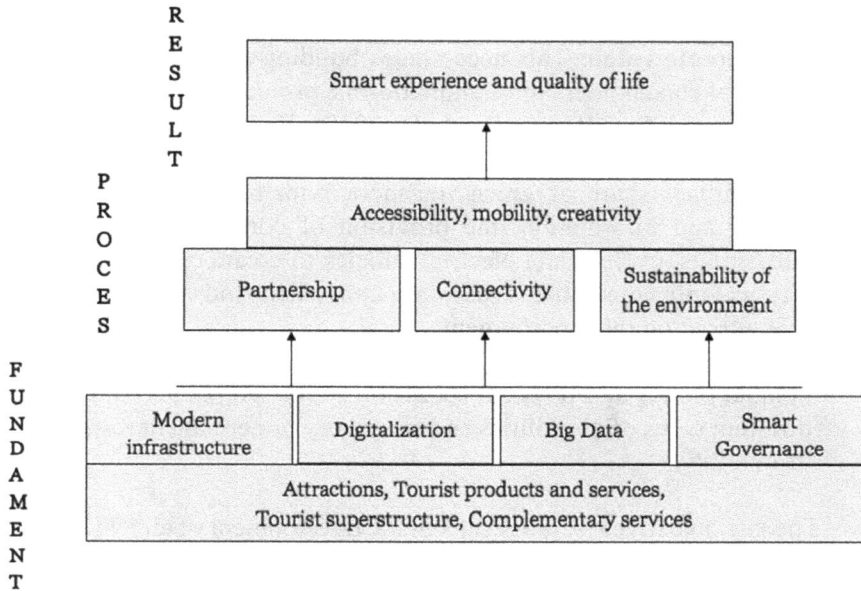

Fig. 1.2.   Smart tourism destination.

Source: Developed by the authors on the basis of Gretzel *et al.* (2015), Buhalis (2016), Perles-Ribes and Ivars-Baidal (2018) and Rafailova (2020).

(4) **Databases:** These are accessible and used by businesses, tourists, the local community and public institutions, formed by many different sources: users, devices, surveys and reviews, official statistics, private and public organizations and contextual information.

(5) **The Six As (Buhalis, 2000; Gajdošík, 2018):** This includes attractions (natural and anthropogenic), accommodation, complementary services (banking, communications), accessibility to destination, attractions, tourist products and services and entertainment.

The key characteristics of an STD — processes that lead to an intelligent tourist experience and a higher quality of life — are as follows:

(1) **Environmental sustainability:** This includes environmental protection and implementation of measures to combat climate change through ecological transport and traffic control, pollution prevention and sustainable waste management and the creation of protected natural parks.

(2) **Cooperation and connectivity between stakeholders in order to co-create value:** This necessitates building and maintaining networks of stakeholders through platforms, projects, centers and hubs.

(3) **Smart mobility (Gretzel *et al.*, 2015; Rafailova, 2020):** This involves intermodality and diversification of transport modes with a predominant share of green transport, with real-time monitoring, control and information, the provision of comfort and safety for pedestrians, cyclists and electric vehicles to create opportunities for physical movement that is easy, fast, convenient and with a less negative impact on the environment.

(4) **Accessibility:** This implies social accessibility for people from vulnerable groups and physical accessibility for seniors and those with different types of disabilities and temporary or permanent restrictions of mobility.

The key objectives — the results of the development of an STD — are as follows:

(1) **An intelligent tourist experience:** This is an experience based on modern ICTs, which enable tourists to receive information in real time according to location and time, to be served quickly, comfortably and easily, as well as to make a constant communication link with a place, locals and the personal social environment.

(2) **The quality of life for the local population:** This includes modern infrastructure, digitalization and smart governance in the destination create opportunities for residents to access quality social services, health care and education.

The analysis of the STD theory and research outlines several directions for further and more in-depth studies related to the development of an STD:

(1) the interrelationship between STD development and sustainable destination development;

(2) the interrelationship between STD and attractiveness and the competitiveness of the destination;

(3) smart business: digitalization and innovation and participation in the industry;

(4) the innovation, knowledge and skills of residents and the role of human resources;

(5) STD development strategies: destination assessment and development models.

The scientific literature does not explain sufficiently clearly the relationship between the characteristics of STDs (in the form of indicators) and indicators of sustainable development in economic, social and environmental terms. The authors' view is that sustainability is seen as an element of STDs, and it should be a context for the development of STDs.

Creating and offering a smart experience is no guarantee that the destination is attractive and competitive above all in the long term. The attractiveness and competitiveness depend not only on the quality and level of attractiveness of the Six As for tourists but also on the sustainability of the environment, atmosphere and attitude of local residents to tourism, digitalization of business and applied innovations (Boes *et al.*, 2016; Koo *et al.*, 2016). There is a need to investigate the relationship between STD characteristics and indicators of the attractiveness and competitiveness of the destination. On the one hand, this means taking into account the role and directions for business development in the destination as smart. On the other hand, it is necessary to examine and take into account the importance of the human factor. The smart development of a destination is unthinkable without the necessary knowledge and skills of people — their abilities to create, implement and use innovations, to be creative and to participate in decision-making processes.

The STD development strategy must be locally driven and based on the specific characteristics of the tourist destination and the achievements in terms of smart development indicators. This requires a toolkit that allows for an assessment of the destination and a strategy model that takes into account the defined state and the specific objectives set according to the situation, trends and desires of stakeholders. In this regard, existing STD models and evaluation methodologies, as well as the relationship between sustainable tourism development and STD development, are explored.

1.2. Models and Methodologies for Research of an STD

Despite the understanding that the future belongs to smart tourist destinations, there have been only a limited number of attempts in the scientific literature to develop a model of STDs. This reveals a greater propensity for descriptive and conceptual research with observational methods.

STD research mainly concerns the definition and characteristics of STDs, tourist experiences (consumer behavior, satisfaction), ICT applications (Internet of Things, Big Data, mobile apps, social media), sustainability (environmental protection) and the smart city. According to Bastidas *et al.* (2020), in the studies on smart tourist destinations, most attention is focused on ICT (40%), followed by tourist experiences (20%), defining STDs (20%) and the topics of sustainability (10%) and smart city (10%).

The theoretical debate in the scientific literature co-exists with policies for the development of smart tourist destinations (Ivars-Baidal *et al.*, 2019). These policies vary from country to country as well as continent. While in Asia (China and South Korea), these are oriented toward creating a technological infrastructure for the development of smart tourism, using the capabilities of ICT for the marketing and management of destinations and tourism resources (Gretzel *et al.*, 2015; Guo *et al.*, 2014; Koo *et al.*, 2016; Li *et al.*, 2017; Wang *et al.*, 2013), in Europe, the initiatives are related to the innovation and competitiveness of destinations through the development of smart end-user applications and specific tourism programs (Segittur, 2015), while cases like Italy are those that bind culture and tourism (Graziano, 2014). In Australia, policies focus mainly on smart governance and the use of open data (Gretzel *et al.*, 2015).

According to Ivars-Baidal *et al.* (2019), SEGITTUR's idea of the STD as a tourism destination that is innovative, sustainable and accessible to everyone, based on an infrastructure of the state-of-the-art technology, increases the quality of the destination experience and improves the quality of life for residents, is a practically perfect idea, but can hardly become a reality.

One of the first models of an STD contributing to a holistic and applied perspective is proposed by Ivars-Baidal *et al.* (2016). This model of Spanish researchers recognizes the favorable role of ICT in the formation of STDs based on strategic and related pre-conditions that determine the capacity for action of a destination and the scope of its strategy. The model is structured according to three interrelated levels (see Fig. A.1 of Appendix A): (1) the strategic-relational level based on governance, characterized by public–private partnership to ensure the sustainability of the destination and an open and collaborative environment of innovation; (2) the instrumental level, based on digital connectivity and sensors, so as to configure an information system for the destination that is essential for decision-making; (3) the application level that allows for the development of smart solutions for marketing and destination management and leads to greater efficiency in communication actions and improvement of the tourist experience.

The model has an important role in the development of STDs. First of all, it presents that for optimal construction of the ICT instrumental level, it is essential to identify the management needs and capacities of a tourist destination. Second, it identifies the need to implement an STD strategy that responds to the local context and how each destination integrates into the global tourism ecosystem. Third, the application of the systems approach highlights how intelligent solutions provide feedback to the main aspects of the strategic-relational level (e.g., innovation or cooperation between stakeholders) and at the instrumental level (increasing the available data on destination information systems). Thus, according to the authors, the development of a smart tourist destination generates synergies that lead to continuous improvement and the creation of a process capable of transforming the management of tourist destinations. Fourth, the model or the "STD approach" according to Ivars-Baidal *et al.* (2016) ensures an increase in the competitiveness of the destination, better satisfaction of demand and the development of new products. In addition, this approach helps increase the expenditures of tourists and develop a public–private partnership. Other advantages are an increase in demand and the emergence of new sources of funding in destinations. According to the authors of the model, it can be assumed that the STD approach improves the efficiency of tourism management and increases the competitiveness of the destination, but does not necessarily lead to an increase in the number of tourists.

Boes *et al.* (2016) offer their own model, which they call the "smart tourism destination framework" (see Fig. A.2 of Appendix A). According to the cited authors, with the rapid development of technology, the ecosystem approach is recognized as appropriate to deal methodologically with the topics of smart cities and STDs. An ecosystem from the point of view of the S-D logic is outlined as a relatively self-contained, self-tuning system of resource-integrated actors, linked through shared institutional logic and mutual value creation through the voluntary exchange of services.

In this sense, the S-D logic explores the interaction between all actors in the ecosystem, the social norms present in it and the reintegration of operand and operant[7] resources for the co-creation of value. Central to the

---

[7]An operand is subject to mathematical or other operations. It is usually expressed in computer programming as constants or variables.

S-D logic is set aside for the purpose of co-creating values inside smart tourist destinations.

The study outlines several baseline elements and interrelationships of the STD model. First, the availability of "hardware" and "software" intelligence, which, according to the cited authors, is critical for the development of ecosystems in smart places. Intelligence develops on revolutionary technologies and innovation, social and human capital and leadership. According to Boes *et al.*, in order to achieve full intelligence in tourist destinations, it is important to understand the relationships between the main components and to facilitate synergy between them. Human capital and people are identified as operational resources and thus as integrators of knowledge and skills in the smart tourist destination ecosystem. It is the human capital that drives innovation, which in turn creates the conditions for the co-creation of value. The role of ICT has been identified as an operand resource.

Second, the cited authors show in the developed model that smart places adopt the structure of an ecosystem based on the logic of the dominance of the value-creation service for all stakeholders who exchange knowledge and skills, which contributes to the success of a smart tourist destination. Third, a smart tourist destination evolves from a destination that develops and refines the pillars of its competitiveness and attractiveness.

The theory presents several methodologies for measuring and evaluating the progress of smart destinations. The indicators they offer are as follows:

- crucial destination planning and management tools often used by public administration and policymakers in the design of measures, actions and plans, according to Ivars-Baidal *et al.* (2021);
- tools essential to monitor the effectiveness and readiness for further advancement of smart destinations, but so far, a unified system is lacking (Alonso Dos Santos *et al.*, 2021).

In this respect, STDs follow the example of smart cities.

In the case of smart cities, indicators have been adopted by many public organizations, including the European Innovation Partnership on Smart Cities and Communities (EIP-SCC). There are also parallel initiatives such as Smart Tourism Capitals in the EU, and indicators may be required to monitor the position of destinations in the indices and track the progress of potential applicants for funding, programs or initiatives.

Komninos *et al.* (2013) believe that, despite increasing research on STDs, it is insufficient to indicate how each city applies this "smart paradigm" to reveal its potential for STDs.

Alfonso Vargas-Sánchez (2016) uses a dual approach for applying complexity theory to STD research, attracting two groups: representatives of scientific thought and professionals at the senior management level in tourism companies and organizations.

The professional approach is based on the following main dimensions:

- *Strategic*: This refers to government commitment and public–private partnerships to ensure sustainable destination management (economic, social and environment) and improve branding management.
- *Operational*: This includes the specific functions of each agent involved in a destination, e.g., institutional support for the implementation of the communication/dissemination task within local society (among citizens, SMEs and institutions) to bridge the technological gap in the digital age.
- *Technological*: This involves designing, integrating and implementing different technologies that can add value and maximize customer satisfaction.
- *Accessibility and information management.*

The academic group also offers four dimensions, but using different denominations. The two approaches are interrelated and can be successfully integrated.

The two groups (academic and professional) identify eight aspects/categories of fields of activity: environment; mobility/transport infrastructures; co-creation of products with tourists; service; information management at the destination; tourist information before travel; tourist information during travel and post-trip tourist information, with the top three priorities being information management at the destination, tourist information before travel and the co-creation of products/services together with tourists: itineraries and destination marketing.

The functions of an STD's governing body are dealt with under three headings: demand-related, supply-side and supply-side connectivity. According to Vargas-Sánchez (2016), a new look for tourism management is also needed, based on four elements: the governing body, technology capital (ICT), human capital and values.

Femenia-Serra and Perea-Medina (2016) propose their methodology based on the qualitative approach by offering a study of numerous case studies (in the case of Malaga, Alicante and Marbella in Spain) to measure how destinations implement smart tourism strategies (see Table A.1 of Appendix A). Their research is based on the understanding of Boes *et al.* (2015) that the case study method is particularly suitable for areas that are undergoing rapid change, such as technology and, in this case, STDs.

According to Femenia-Serra and Perea-Medina (2016), the study of multiple case studies allows for the comparison of similar destinations, which at the same time have different features. This allows opportunities for benchmarking (i.e., a business technique that consists of using a benchmark) to help destination marketing organizations.

To define the main categories, Femenia-Serra and Perea-Medina (2016) used those defined by SEGITTUR (2015) — sustainability, accessibility, technology and innovation — as well as the developed system model by Ivars *et al.* (2016) (see Fig. A.1 of Appendix A) — governance, sustainability, innovation, connectivity and sensor networks and information systems.

Possible estimates for each category include missing (deficient); unfavorable; acceptable (acceptable); favorable and advanced (advanced).

According to Femenia-Serra and Perea-Medina (2016), this methodology can be considered a first step to assessing a smart tourist destination and creating an appropriate framework for indicator development. It is recommended that the qualitative analysis be supplemented by a quantitative survey to obtain a complete picture of the situation of each destination, the purpose of which is to be "intelligent."

Ivars-Baidal *et al.*'s Valencian network of smart destination indicators (2021) is based on the Smart Cities Assessment of Giffinger *et al.* (2007), sustainable tourism destinations (European Commission, 2016) and the Tourism Competitiveness Assessment (WEF, 2019). This model is one of the newest and most applicable at the regional level.

Ivars-Baidal *et al.* (2021) believe that so far there is no system of indicators to measure how destinations are progressing in terms of expectations and goals in the direction of smart tourism. The cited authors are developing such a system of indicators to reveal the current situation in specific Spanish tourist destinations regarding the extent of their progress toward becoming smart destinations.

Ivars-Baidal *et al.* distinguish between case studies using other destination assessment methods or multi-destination comparison, as abound in the scientific literature and research on the identification and use of smart-tourism indicators. The logic for creating indicators is similar in most cases: building on a model or framework for smart cities.

The model of Ivars-Baidal *et al.* (see Fig. A.1 of Appendix A) discussed above was subsequently used to determine the indicators that provide fundamental information and examples of different types of dimensions.

This holistic model conceptualizes a smart destination, structured in three interrelated levels with their respective principles. At the strategic-relational level, a smart destination is based on management, public–private partnership and coordination in the administration in order to achieve sustainable tourism development, an innovative environment and an accessible tourist territory for all (see Fig. A.3 of Appendix A).

According to the indicator "governance," there are 10 indicators, some of which are the following: (1) implementation of a strategic plan for tourism, (2) coordination mechanisms between the departments of the local administration for the development of smart-destination projects, (3) availability of a coordinator of the smart destination, and (4) mechanisms to facilitate public–private partnership.

According to the indicator "sustainability," 15 indicators are applied, such as the following: (1) implementation of urban planning rules consistent with the principles of sustainability, (2) improvement of energy-efficiency strategies (public lighting) and (3) efficiency in water supply, wastewater treatment and reuse.

At the first level, there are six indicators on the third indicator "innovation" and also on the fourth indicator "accessibility."

The second, instrumental level is built on digital connectivity (with five indicators), sensory and Big Data, on which information and intelligent systems rely (with a total of 17 indicators). This system facilitates the interaction between the physical and digital worlds, a key feature of smart tourism. The third, application level generates intelligent solutions for marketing and destination management (eight indicators) as well as for improving the tourist experience (with five indicators).

The set of indicators has been pre-tested in a small number of destinations. Indicators for adequate and updated information that is not available have been rejected, resulting in a set of 72 indicators distributed in three levels and nine criteria (see Fig. A.3 of Appendix A). The final set of indicators was used to evaluate 18 destinations from the Valencia region (including the city of Valencia), with 15 destinations being referred to level three (most advanced STD projects) of INVAT. TUR[8] and the other three were referred to level two, which approximates the highest level.

Due to the geographical and economic structure of the Valencia region, there are differences between destinations in terms of the specialization of their local economy in tourism. This gives the authors a reason to divide the destinations into coastal and inland, by population, by number of accommodation establishments and by number of beds. Cluster analysis techniques are also used to divide the destinations into the following: cities with a medium-to-low tourism function; intermediate cities with a medium-to-low tourism function; mixed destination; tourist destinations where local residents also live (residential tourist destination) and destinations with only a hotel destination and a small agricultural municipality with limited tourist activity (agrarian small municipality with limited tourism activity).

Indicators are necessary to measure the effectiveness of management and the implementation of objectives, which in the case of STDs are determined by planning tools. Indicators are also indispensable in the analysis of new governance models. According to Ivars-Baidal *et al.* (2021), the system of smart destination indicators developed by them, with the institutions (INVAT. TUR) and local municipalities, allows for the identification of the strongest and weakest points of the analyzed destinations.

The smart tourism development assessment system of Ivars-Baidal *et al.* (2021) has so far only been used in selected regions of Spain. The authors propose applying the model to other countries in order to develop and improve the system of indicators.

Maráková *et al.* (2022) tested the model's performance in Banská Bystrica, Slovakia. The results of the study contribute to assessing the adequacy of the indicator system at the regional level. Maráková *et al.* (2022) improved the system of indicators. After preliminary tests, the criteria were refined, reformulated and adapted, reducing them to 70 indicators due to the inability to obtain statistics for the other two. The authors proved that the model is applicable to all regional governance organizations in Slovakia, giving a specific answer in which dimensions the region of Banská Bystrica lags behind or progresses.

---

[8] The Valencian Institute for Tourism Technology, conceived as a meeting platform for all agents in the tourism sector, represents one of the main axes for improving the competitiveness and sustainability of the Valencian community tourism model.

Therefore, as the potential for the development of a smart destination may vary from region to region, a different set of indicators should be used depending on which are available and correspond to the specifics of the destination. The methodology is constantly improving, taking into account the specifics of the dynamic development of smart tourist destinations in recent years.

1.3. Sustainable Development of Tourism, Tourist Destinations and Their Management

Tourism today is characterized by positive and negative processes and impacts on the environment — natural and socioeconomic (O'Reilly, 1986; Valentine, 1993; Kercher, 1993; Murphy, 1994; Cooper and Archer, 1994; Butler, 1993; Fletcher *et al.*, 1994; Urry, 1994; Rakadzhiyska, 1997; Vodenska, 2001; Marinov, 1998, 2003; WTO). They can be classified into the strands of generally significant environmental, social and economic problems that are interrelated, with long-term effects, which are often global. Their complex solution and prevention necessitate a systematic and integrative approach to the development of society, such as sustainable development. Numerous modern scientific and practical studies and theoretical developments in the field of tourism are aimed at analyzing and assessing the negative and positive influences of tourism — the direct or indirect consequences of its development, with immediate or cumulative long-term effects.

The review and assessment of the environmental impacts of tourism — natural, economic and social — in the scientific literature (Marinov *et al.*, 2022; Hunter and Green, 1995; Bieger, 1997; Holden, 2001; Bosselman *et al.*, 1999; Pearce, 1993; Dyer *et al.*, 2003; Tosun, 2001; Ryan, 2002; Vodenska, 2001; EC, 2010) lead to the following conclusions:

*   Tourism can destroy the resources, factors and conditions that determine its development, economic efficiency and role in society.
*   The negative and positive results from the influence of tourism are interrelated and manifest in various directions. These consequences can be observed mostly at the local level. Their management requires an integrative and specific approach in terms of objectives, tools and mechanisms, indicators of analysis and control and ways

of making a decision. However, this can only be carried out in accordance with regional and national development strategies and plans.

- Improper management of the sector causes negative impacts that can have an irreversible and long-term effect not only for tourism but also for the local population, other business activities, political and social life, and future generations of both tourists and locals.

The results of the studies confirm the need for a change in the policy, strategy and method of tourism development to obey the principles of sustainable development. In response to these challenges and problems, the concept of sustainable tourism development has been created. At the same time, a number of guiding rules for its implementation have been introduced, and established mechanisms, tools and strategic approaches have been established, through which its objectives can be achieved.

The concept of sustainable tourism development is a private theoretical model for the transformation and functioning of tourism in the direction of solving and preventing the problems it itself raises and providing a contribution to the achievement of the sustainable development of society. It covers specific practical programs and rules for the management and development of tourism, which lead to positive results simultaneously in environmental, economic, social and political aspects.

For the purposes of this monograph, the authors accept the concept of sustainable tourism development, which is based on the principles of sustainable development, because

> sustainable development in the context of tourism should be considered as: tourism that develops and maintains itself in a given area (community, natural environment) in a way and at a scale through which it is effective for a long period of time and does not lead to degradation or change of the environment (physical and human) in which it exists, to a level above which the successful deployment and vitality of other activities and processes is hindered. (Butler, 1993; Lew, 1998).

In this way, the following are clearly defined:

- The goals that can be achieved are the long-term sustainability of the tourism business and maintaining the revenues from it, sustaining the quality of the environment in order to attract tourists and contributing to improving the quality of life and the living environment.

- The effective exploitation and conservation of capital needed by tourism as well as the local population, the nation and the global community is important for environmental sustainability.
- The needs and interests that will be met and protected are of not only today's key actors and social groups but also future generations and socially disadvantaged groups that may be directly or indirectly affected by the development or limitation of tourism activities.

The concept of sustainable tourism development is most fully and summarily expressed through the definition derived from the Canada Tourism Board, the World Tourism Organization (1990) and Inskeep (1991), which states as follows:

> The sustainable development of tourism meets the needs of current tourists and local populations while protecting and preserving opportunities for future periods of time, i.e. it does not prevent the next generation from satisfying its needs and desires by managing all resources in such a way as to satisfy economic, social and aesthetic needs, while maintaining cultural identity and integrity, essential ecological processes, biodiversity and life-defining systems.

The definition clearly states the objectives of sustainable development of tourism as well as the fundamental principles:

- for environmental sustainability;
- for the use of natural resources and social values while preserving the indispensable elements of capital and its total value;
- to preserve and enhance the welfare that the future generation (tourists, local population and business) will inherit;
- to achieve a balance between the interests of key actors in tourism and all social groups.

The concept of sustainable tourism development is integrative in nature, which implies simultaneously achieving positive results in the field of environmental protection, the social and economic sphere and the construction of civil society.

Achieving the sustainable development of tourism according to the established principles requires developing and implementing strategies and plans primarily at the level of a tourist destination.

A tourist destination is a category that is considered in different aspects as follows: a geographical area visited by tourists with different motives and goals, a complex of tourist services, goods and infrastructure perceived as a whole and a concentration of tourist services and goods managed by an organization bringing together, most often, representatives of business and local authorities:

> Destinations are places to which people travel and in which they choose to stay for a while in order to "experience" certain characteristics and attractions. (Leiper, 1995).

> Destinations are a concentration of base and services created to satisfy tourists' needs. (Cooper, 1993).

For the purposes of this study, the definition of Buhalis (2000) is adopted: "A tourist destination can be seen as a defined geographical area that is perceived by visitors as a specific entity and that has a political and legal organizational structure for tourism marketing and planning." It follows from this that a tourism destination:

- includes a set of tourist services, goods and infrastructure and an anthropogenic and natural environment, an atmosphere that plays a role in the experience of tourists, for the local community and for other associated communities;
- has a territorial scope and structure in the context of its management.

The concept of sustainable tourism development must be tailored to specific characteristics and problems at the local level first (Cooper, 1993; Butler *et al.*, 2002; Murphy, 2003; WTO, 2005; EC, 2010). This is because the negative influences and consequences of a lack of integrative and long-term planning of tourism (OECD, 1981) appear locally first. Moreover, the socioeconomic and natural conditions in which a sustainable tourism development policy is pursued differ in each tourism destination (Cooper and Archer, 1994; Pearce, 1994; Horn and Simmons, 2000; Tosun, 2001). It is its territorial spatial and regional plan that outlines the specific boundaries and approaches to the use of natural resources and sociocultural wealth. A tourism destination is characterized by distinctive conditions and factors that are a prerequisite, barrier or opportunity for its sustainable development.

Solutions to the specific environmental and social problems and those for improving the quality of life could not be found if only individual participants in the chain of creation of the tourist product or related institutions, organizations, etc., follow the principles of sustainability. All of them should be involved in a concept for the sustainable development of a tourism destination. This concept needs to be specific and to express the sustainable development of tourism in functional, territorial and temporal aspects. To do this, general guiding rules should be followed (Goodall and Stabler, 1997; Inskeep, 1991; UNEP; WTO), namely the principles of sustainable development of the tourism destination. They can be summarized in four groups (see Table 1.6), each of which derives from the relevant general principles of sustainable tourism development:

(1) preserving and improving the quality of the environment and the living environment in the tourism destination;
(2) preserving and improving the quality of the tourist product — experiences, services and infrastructure;
(3) satisfaction and respect for the interests of all key actors and social groups;
(4) integration of the local population/residents and companies in the process of decision-making and distribution of created goods directly or indirectly from tourism.

In conclusion, the concept of the sustainable development of a tourism destination is a theoretical construction built on the principles of the sustainable development of a tourist destination. These principles are the foundation of defining destination development goals and criteria for measuring achievements in sustainable development.

The review of the scientific literature, policy programs and implemented plans and instruments sets out two main directions for measuring sustainable development:

(1) impact of tourism on the development of a destination in environmental, economic, social and political aspects;
(2) implementation of strategy, plans and programs for the sustainable development of a destination.

Table 1.6.   Principles for the sustainable development of a tourist destination.

Principles of sustainable development of a tourism destination	Principles for the sustainable development of tourism		Specific performance in a tourism destination and the role of tourism in sustainable development
	Basic	**Supportive**	
(1) Preserving and improving the quality of the environment and the living environment[a] in the tourist destination	Principles for the ecologically sustainable development of tourism	Principle of preserving the value and nature of total capital Principle of increasing wellbeing	1.1. Contribution to preserving the value of capital of society and the economic sphere at the local level 1.2. Contribution to the conservation and protection of non-renewable and renewable natural resources at the local level 1.3. Contribution to the conservation and protection of natural factors and their properties 1.4. Contribution to the storage and protection of local ecosystems and landscape
(2) Preserving and improving the quality of the tourist product — experiences, services and infrastructure	Principles for economically sustainable tourism development	The principle is for the protection and conservation of natural resources, functions and landscape	2.1. Preserving the value of tourism capital — natural capital (especially irreplaceable)

			The principle of protection and preservation of socio-cultural wealth	2.2.	Preserving the value of the capital of tourism — human-made in the tourism destination
				2.3.	Preservation of the value of the capital of tourism — local socio-cultural capital
			The principle of increasing the quality of experience of tourists	2.4.	Preserving the atmosphere and beauty of the destination
(3)	Meeting the interests of all key actors in the value chain and societal groups	The principle for sustainable development of tourism in the social aspect	The principle of preserving and enhancing wellbeing	3.1.	Contribution to improving the quality of life of the local community and tourists
				3.2.	Preservation and growth of the value created in tourism and the revenues of the local business and its partners
				3.3.	Contribution to the regional value/ welfare process
				3.4.	Contribution to the protection of the rights of local ethnic groups and minorities
(4)	Integration of the local population and companies in the process of decision-making and distribution of the created goods	Principles for the sustainable development of tourism in the institutional aspect	The principle for establishing and affirming legal, legislative and regulatory conditions for sustainable development in the social aspect.	4.1.	Contribution to the establishment of legal, political and institutional conditions for the participation of local residents in the decision-making process
				4.2.	Contribution to the fulfillment of local specific requirements and needs

Note: [a] Complex category, understood as a set of vital conditions and conditions for the implementation of social activities, evaluated mainly by representatives of the local population.

Source: Elaborated by the authors from Rafailova (2005).

In the first direction are the indicators "driving forces" representing activities and processes of tourism influencing the environment:

(1) **Stress/pressure:** Number of tourists per year/season — number of tourists in a destination — total, foreigners, in the high season and compared to other seasons, and rate of change over the years, including a forecast.

(2) **Intensity of demand-side use of a tourist destination:**
  (2.1) Number of tourists and residents per unit area (km^2), number of tourists per unit area, number of tourists and local residents per unit area compared to the value of local residents per unit area as well as the rate of change over the years;
  (2.2) *Nights per unit area*: These include average nights per unit area and comparison to similar destinations, along with national and European averages.

(3) **Intensity of supply-side use of the tourist destination:**
  (3.1) *Number of beds per hundred people*: tourists and residents;
  (3.2) *Number of beds and accommodation establishments per unit area* (km^2): The values of this indicator can be presented compared to the rest of the administrative area (or other destination) and other similar destinations in the country and Europe or with an average value in the country.
  (3.3) *Areas used by tourism*: These include the size of built-up areas with hotels, with coastal restaurants and facilities, number of restaurants, entertainment, accommodation along the coast in parks and forest areas. The values under this indicator may, if possible, be presented in comparison with the approved territorial development and cadastral plan for the municipality or the respective administrative region.

In the second direction are indicators presenting achievements: measures and means to protect the environment and achieve environmental sustainability, indicators of progress in achieving sustainable economic development (including building an economy based on knowledge and information, relationship of economic policy with principles of environmental and social sustainability), indicators presenting progress in achieving social justice and integration, including measures and means for preservation protection and development of sociocultural wealth and historical heritage, and opportunities for the development of the individual and the community. In this aspect, the indicators of sustainable development of a tourist destination are divided into four groups (see Tables 1.7 and 1.8):

(1) environmental sustainability;
(2) economic sustainability;
(3) social sustainability;
(4) effective destination management.

Table 1.7. Indicators for the sustainable development of a tourist destination: Environmental sustainability.

Environmental sustainability	Indicator	Measurement indicators
Climate change	Use of alternative energy in	Share of tourism companies
Air quality	the destination	using alternative energy
Water resources management	Managing the impact of climate change	Share of tourism companies implementing schemes to
Waste management	Use of pedestrian and bicycle	reduce the impact of
Sewage management	transport	climate change
Noise pollution	Possibility to use alternative	Share of tourist products or
Urbanization	modes of transport — on	trips of tourists based on
Habitat protection and biodiversity	foot, bike, scooters, etc. Systems for efficient use of	pedestrian/bicycle transport Availability of pedestrian
General measures for the protection of the environment	water for tourism business purposes	zones, alleys, places for renting a bicycle or scooter,
	Waste-management systems used in tourism and the	etc., — length, number Share of tourism
	destination	organizations implementing
	Ratio between biological and	controlled water use
	mechanical treatment by	systems
	treatment plants with	Share of tourism companies
	receiver water basin used	with separate waste
	for recreation purposes,	collection/recycling
	presence of double	systems
	purification	
	Use of electric transport	

(Continued)

Table 1.7.   (*Continued*)

Environmental sustainability	Indicator	Measurement indicators
	Integrated urban planning, including integrated coastal zone management Protected natural areas Environmental management	Number of tourist companies connected with treatment plants for biological wastewater treatment/own wastewater treatment plants/double wastewater treatment
		Share of tourism companies using electric transport vehicles
		Construction restrictions — architecture, size, means used, zones
		Size of protected areas per capita or share of total areas
		Share of tourism companies actively involved in habitat and biodiversity protection
		Share of tourist companies holding a certificate (EMAS, ISO 14001, Ecolabel) for an environmentally oriented company

Source: Developed by the authors from Kazandjieva *et al.* (2022), Miller (2001) and EU (2006, 2016).

Based on a theoretical review and analysis, the authors determine the main thesis of the study as the development of an STD 2030 is possible if it is carried out in close interrelation with sustainable development and is socially oriented.

The first hypothesis to prove the main thesis is as follows:

H1: There is a need to develop a smart destination model based on the principles and goals of sustainable development that is human-centered.

Table 1.8. Indicators for the sustainable development of a tourist destination: Economic and social sustainability and effective destination management.

Economic sustainability	Indicators	Measurement indicators
Energy efficiency and sustainability Tourism contribution to economic development Transport	Energy used vs. value created Tourist demand Revenues Employment Business performance Transport modality Access to the destination	Consumed kW per unit of product/night, stay or by accommodation, etc. Number of tourists per month, year GDP share of tourism in percentage Average arrivals of tourists per day Share of tourism employees in relation to total employees Occupancy rate of accommodation Interrelationship between different modes of transport Ratio between tourists using different modes of transport for access to and in the destination (air, car, water, public, bicycle, etc.)
Social sustainability	**Indicators**	**Measurement indicators**
Social integration Preserved identity	Access for representatives of socially vulnerable groups Preserved tangible and intangible cultural heritage, architectural design, landscape	Facilities for persons with limited mobility in accommodation, tourist attractions and places — number or share of places with access in relation to the total number Share of tourism companies participating in initiatives for the preservation of cultural heritage and/or offering related products Share of visits to cultural heritage sites in relation to the total number of visits to the destination

(Continued)

Table 1.8.   *(Continued)*

Effective destination management	Indicators	Measurement indicators
Satisfaction of tourists Participation in decision-making Image of the destination	Overall satisfaction and satisfaction with infrastructure, food and drink, events, entertainment and tourism products Representation of tourism business in governing bodies responsible for decision-making for tourism and destination development Associations, emotions related to the destination Image of the destination	Share of tourists satisfied with destination relative to total number of tourists Share of tourists making repeated visits to the destination Commissions responsible for tourism policy with the participation of representatives of tourism Percentage participation of tourism representatives in commissions responsible for economic development of the destination (transport, urbanization, infrastructure) Degree of positivity, character of associations and emotions, including in the media

Source: Developed by the authors from Kazandjieva *et al.* (2022), Miller (2001) and EU (2006, 2016).

1.4. Smart and Sustainable Tourism Destination Model

1.4.1 The relationship between sustainability and STDs

The interrelationship between the construction of STDs and sustainable development is accepted as a mandatory condition for the development of a destination as smart according to the fundamental definition of SEGITUR. But, both in the definition itself and in numerous studies, the nature of this interrelationship is not explained, but is rather perceived as natural in the sense that sustainability is a consequence of an application of the concept of STDs (Vargas-Sánchez *et al.*, 2018) or is understood by presumption (González-Reverte, 2019), i.e., a destination is assumed to be smart if it is sustainable. Politicians and researchers unite (First WTO Smart Destinations Conference, 2017) around the conclusion that the intelligent approach to tourism management is a driver for sustainable development in terms of the complexity of long-term decisions made using technology and Big Data. This, however, turns out to be a theoretical explanation (Perles-Ribes and Ivars-Baidal, 2018) rather than a practical one because STDs' development strategies are rather aimed at increasing the attractiveness of the destination and the satisfaction of tourists. On the other hand, research that examines the relationship between sustainable development and STDs focuses mainly on environmental and economic sustainability.

Environmental sustainability and the development of STDs are presented primarily as a causal link between the application of technologies and the effective management of environmental impacts, including tourism (Perles-Ribes and Baidal, 2018), to increase energy efficiency and environmental friendliness of transport and regulate tourism and resource efficiency in the sector (Vargas-Sánchez *et al.*, 2018; Perles-Ribes and Baidal, 2018).

The interrelationship between economic sustainability and STD development is considered in terms of creating the right infrastructure and environment for developing a competitive business, building networks and a database allowing for better sustainability of the economy and tourism in terms of changes in resource prices, political conflicts, emergencies such as pandemics and natural disasters.

Some STD models have been proposed in the scientific literature, in which the relationship with sustainable destination development is presented:

- model of Perles-Ribes and Ivars-Baidal (2018);
- models for a sustainable smart tourist destination of Shafiee *et al.* (2019).

According to Ivars-Baidal and Perles-Ribes (2018), the concepts of sustainability and smartness share many common elements, but a destination cannot be considered smart if it is not sustainable. They enrich the content of the previously developed system model (see Fig. A.1 of Appendix A) by formulating the common elements between STDs and sustainability. Ivars-Baidal and Perles-Ribes (2018) even point out that there are proposals to use the term "smart sustainable cities" as a more accurate term, as its use is growing in both cities and tourism destinations, in order to better highlight the sustainability dimension. In this regard, the authors propose a synergistic model between intelligence and sustainability (see Fig. A.4 of Appendix A).

The model identifies the main aspects of intelligence and sustainability and the possible synergies between the two concepts, presenting the relationship between them through certain technologies. It sets out six main functions of sustainable-tourism policy: planning in a long-term perspective; creating a scenario; more efficient use of resources; real-time governance; public–private partnership and open innovation; greater transparency and personalization of tourism services.

The use of technological advances leads to a number of positive actions that help make a smart destination sustainable. These include the more optimal utilization of natural resources, an effective system for monitoring and decision-making in real time, more stable interconnections between the private and public sectors, satisfied tourists and better marketing of the destination. These improvements can attract more tourists but with less intensive resource consumption and a less negative impact on the environment.

Shafee *et al.* (2019), for example, proposed a model for STDs where sustainable development is at the center. The cited authors use the grounded theory[9] method as a framework for their analysis. By creating their model (see Fig. A.5 of Appendix A), Shafiee *et al.* concluded that when causal conditions in society utilize modern technologies for the economic and social development of tourism destinations, these conditions become factors for the successful development of smart destinations and are also accepted as facilitating prerequisites and incentives influencing the dynamics of this development.

The cited authors divide the factors that influence the process into three groups. The first comprises "existing circumstances," which include the achievements of technology (social networks, Big Data), the emergence of smart cities, global trends and economic and social development. The second group consists of context conditions, such as economic and financial, technical and infrastructural and natural and cultural factors. The third group consists of modeling factors related to support from local governments. The process involved in making a destination smart is

---

[9] Grounded theory is a systematic methodology that is applied to qualitative research. The methodology involves the construction of hypotheses and theories through data collection and analysis.

complex and long. Based on the model, it can be concluded that when using modern technologies for the economic and social development of tourism destinations, they become factors for the successful development of smart destinations. The presented conditions stimulate environmental, economic, social and technical actions. These, in turn, lead to improving the tourist experience, improving the management of natural resources, improving the quality of life of the local population, etc.

Theoretically, smart destination models do not set out the relationship between digitalization, smart governance, improving the quality of life and experience of tourists and the environmental, social and economic sustainability of the environment clearly enough. Despite the existing toolkit for assessing a destination as sustainable and the concepts for defining it as smart, it is necessary to link these two approaches in a common model.

The conclusions of the review of STD studies and models are as follows:

- Sustainable development is taken as a characteristic, a consequence or a condition of STDs, but is not necessarily tied to the destination development objectives and priorities as an STD.
- There is a strong focus on environmental sustainability or enabling businesses to be competitive through digitalization without taking full account of social sustainability, with the exception of creating conditions for serving socially vulnerable groups. The development of STDs is not sufficiently linked to the preservation of cultural values and intangible cultural heritage, the stimulation of creativity and small business or the development of human resources, their knowledge and competencies and support for an active citizenship.
- The presented models concentrate mainly on the process of digitalization and sustainable development. The human factor (people) is identified as an operational resource and is not covered in enough depth. This could be seen as a weakness or underestimation of the role of human capital, as all scientific and technological achievements should focus on the wellbeing of both the local population and tourist visitors of the destination, and increasing its attractiveness, otherwise it becomes an end in itself. The quality of the tourist experience is the goal of the development of STDs, but it is also necessary to take into account both the local context and the motives of tourists for visiting the destination, e.g., relaxation, entertainment, spiritual enrichment,

communication and authentic experience, which do not always require the use of digital technologies. They are achieved in an environment that is socially and environmentally sustainable as well as accessible and secure (Vargas-Sánchez *et al.*, 2018), with preserved values and resources, an authentic appearance and identity. The quality of life of the local people is also an objective of the development of STDs, but it is necessary to ensure their participation in the development of the destination and to achieve satisfaction with the process of co-creation of the tourist product. If a destination wants to develop as a smart one, it is imperative that local citizens and tourists are at the core of the whole process.

• Operational models of an STD are needed to provide a link between the scientific aspect of the concept, management in terms of policy and decisions and the specific context of the destination.

Taking into account the above conclusions, the authors proposed a model for a smart and sustainable destination, in which the concept of smart development is interrelated with sustainable development by applying a human-centric approach.

1.4.1.1　*Model for a smart and sustainable tourism destination*

The construction of the model for a smart and sustainable tourism destination is based on the existing approaches to the development of an STD by linking its goals with the principles of sustainable development and the social aspect of their achievement.

The model presents a combination of determinants and mechanisms of an STD, oriented toward its sustainable development and the quality of experience, life and the tourist environment in a smart ecosystem with a focus on social and human factors (see Fig. 1.3).

The goal of developing a smart and sustainable tourist destination is to achieve the satisfaction and wellbeing of tourists and locals at the same time while preserving the public interest in territorial and global aspects. It integrates the goal of a valuable experience for tourists and the quality of life of the local population with the social orientation and sustainability of the environment — the development of an attractive tourism destination that is also a qualitatively urbanized place to live, work and prosper for a vital local community, sharing values, ecologically minded and with active measures to combat climate change.

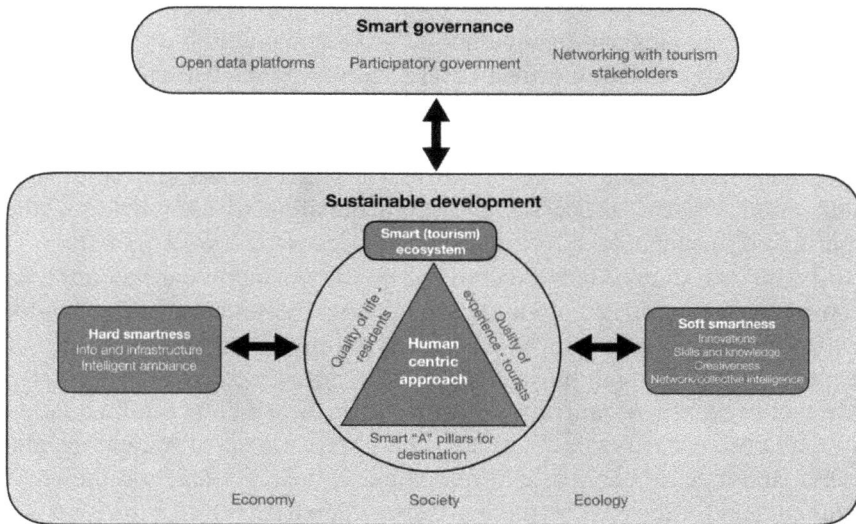

Fig. 1.3. Model for a smart and sustainable tourism destination.

Source: Developed by the authors based on Gretzel *et al.* (2015), Ivars-Baidal *et al.* (2021) and Boes *et al.* (2016).

The essence of the model is determined by smart governance, which initiates, supports and develops a smart ecosystem by combining "hardware" with "software" smartness in a sustainable environment in an ecological, economic and social, long-term foreseeable aspect, leading to an increase in the quality of the experience of tourists, quality of life and attractiveness of the destination.

Smart governance is participatory governance that creates and shares information and resources and uses Big Data and smart technologies to make long-term and complex decisions, with open platforms and an approach of broad access to innovations to bring together and involve stakeholders in the planning, monitoring and control processes. Smart local governments, together with residents, and educational and scientific organizations, develop, support and regulate "hardware" and "software" smartness in the context of sustainable development, which guarantees the strategic aspect of policy.

Sustainable development is defined by the development of a "green" and circular economy, an attractive business environment, the preservation and enrichment of local identity and the achievement of social integration.

Smart governance supports — directly through investments, public–private partnerships and/or regulation — the construction of "hardware" smartness, which covers infrastructure with integrated digital technologies and developed information systems. Investments in smart technologies and their application are oriented toward creating an ambient intelligence environment (Buhalis, 2020), which allows and stimulates innovation, creative processes and the acquisition of knowledge in the context of sustainable development. This leads to the construction of "software" smartness. It is determined by the development and application of innovative ideas, products, technologies, companies and a combination thereof, contemporary skills and competencies and creativity of residents, tourists and businesses. Smart infrastructure is a digitized (Gretzel *et al.*, 2016) and "green" infrastructure, which is the foundation of competitive business activity and the improvement of attractions and tourist services for economic development with less impact on the environment and for the development of partnership networks for the joint use of resources, knowledge and information. Its construction is possible from human resources with digital competencies, motivated to apply innovative approaches and "green" knowledge. Drivers of the further development of a smart environment are clusters and centers for innovation and creative entrepreneurship and business.

A smart ecosystem is formed as a result of the interaction between "hardware" and "software" smartness (Boes *et al.*, 2016) in the context of sustainable development. This ecosystem allows stakeholders to participate in the development of the destination, to be informationally networked and to work collaboratively, with the pillars of the destination's attractiveness developed and managed in a smart and sustainable way. The experience of tourists is valuable — desired and preferred thanks to mobility, the ability to use personalized services and information in real time, to interact with the characteristics of the destination freely, but also to realize the goals of the trip in a sustainable environment, without being entirely dependent on mobile devices and digital technologies, and to participate in building its image and popularity. Local residents enjoy a better quality of life thanks to the smart environment and infrastructure, as well as the access provided to participation in decision-making related to the development of a destination. This ecosystem empowers and encourages tourists and residents to create a collaborative tourism product, share values and have experiences that bring satisfaction to all. At the same time, the attractiveness of a destination is protected and improved, and stakeholders such as tourism businesses, public institutions and NGOs enjoy prosperity.

The role of the model is for the smart and sustainable destination to exist in two main directions:

- assessment of a destination's status as an STD;
- developing a strategy for the smart and sustainable development of a destination.

The implementation of the concept of a smart tourist destination requires, in the first place, an assessment of the status and achievements of the destination against a certain model and existing good practices. This opens the possibility to make strategic decisions, taking into account the local context, for changes and policies regarding the determinants and mechanisms of STDs and sustainable development, as well as to define destination-appropriate goals.

For the assessment of destination status, it is necessary to use indicators that take into account the smartness, sustainable development and social orientation of the destination development. They can be classified based on the main elements and interrelationships between them in the presented model as follows:

(1) **Smart governance:** This includes indicators for digitalization and orientation toward sustainable development and democracy.
(2) **Sustainable development of the destination:** This includes indicators of the development of tourism in the destination in environmental, economic and social aspects.
(3) **"Hardware" smartness:** This includes indicators of digitalization and of contributing to the infrastructure and environment for environmental and economic sustainability and with social orientation.
(4) **"Software" smartness:** This includes indicators of creativity, community intelligence and innovation in the context of sustainable development.
(5) **Smart ecosystem:** This includes indicators of connectivity, sustainable mobility for residents and tourists, socially and environmentally significant accessibility and attractiveness of the destination and competitive business.

Chapter 2

Research and Analysis of a Smart and Sustainable Tourism Destination (Municipality of Varna)

2.1. Methodology of the Study

2.1.1 Description of the methodology

The methodology is based on the author's model of an intelligent and sustainable tourist destination. The main hypothesis is as follows:

> **H2:** The development of a socially oriented strategy for a smart and sustainable tourism destination requires the use of a methodology to assess the level of its smart and sustainable development.

The first stage is the identification of the specific characteristics of the research object (Municipality of Varna). Based on previous research, a survey among important groups in the destination — experts, tourists and citizens — is conducted. Derived from the results, an analysis of the development of Varna as an SSTD is made (Fig. 2.1).

2.1.2 Methodology for assessing a smart and sustainable tourism destination (SSTD)

The methodology for assessing a smart and sustainable tourist destination is based on the SSTD model (see Chapter 1, Fig. 1.3). Similar

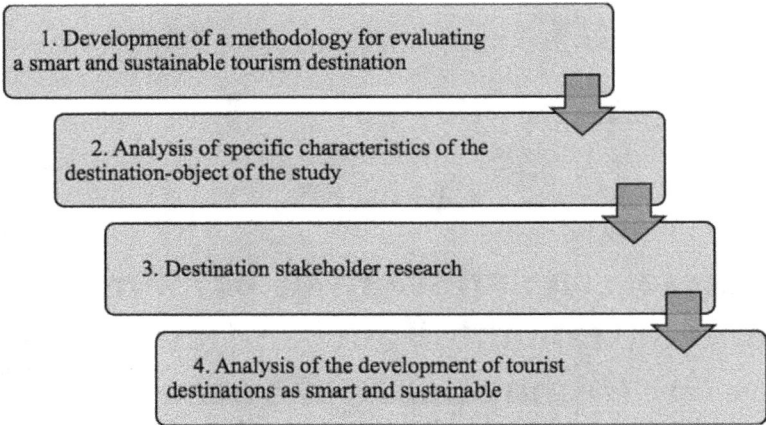

Fig. 2.1.   Methodology of the study.

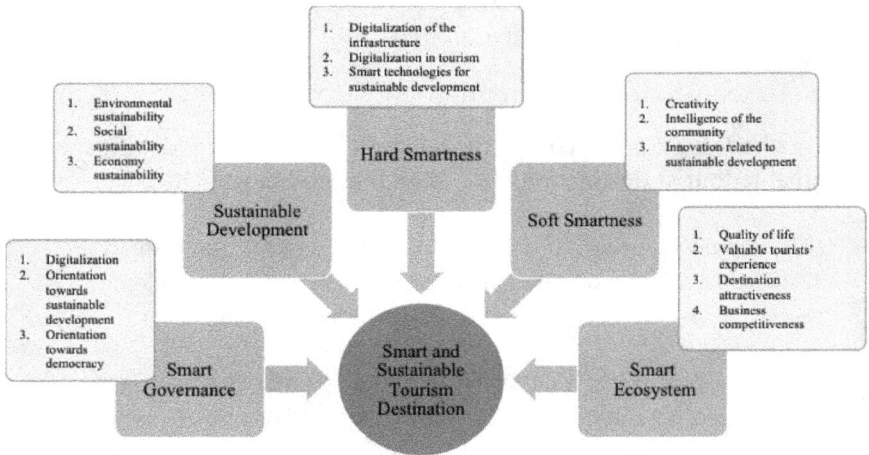

Fig. 2.2.   Methodology for assessing a SSTD.

methodologies are applied in tourist destinations in Spain and Slovakia, as well as a methodology for assessing the sustainable development of a tourist destination and ranking of smart-tourist capitals by the European Commission. It consists of five dimensions (see Fig. 2.2) corresponding to the presented SSTD model, each of which is separated into three groups of indicators, with the exception of the "Smart Ecosystem," to which an additional group is added. In "Smart Governance" and "Sustainable

Development of the Tourist Destination," the groups are divided into three subgroups of indicators. All sectors in software intelligence, hardware intelligence and smart ecosystem include one to four indicators.

2.1.2.1 *Smart governance*

Digitalization

(1) E-government — accessible online information systems and registers for administrative services, e.g. payment of taxes and fees, obtaining permits and licenses, and applying for public funding.
(2) Platforms for management and cooperation with stakeholders — sharing information and open dialog, support partnership and joint projects.
(3) Open database — data regarding tourism, ecology and climate change, transport, demography and economy in the destination, from a variety of sources and accessible in real time.

Orientation toward sustainable development

(1) Strategies and plans for sustainable development of tourism and the destination — responsibility of local authorities and discussions with all parties involved.
(2) Strategies and plans for smart development/infrastructure, digitalization/related to sustainable development — integrated urban plans and strategies for digitalization and innovation. These are the responsibility of local authorities but should be developed with the broad participation of stakeholders and other parties involved.
(3) Monitoring the development of a tourist destination with indicators taking into account sustainable development — they have to be institutionalized and regularly implemented with publicly available results.

Orientation toward democracy

(1) Mechanisms for the cooperation and participation of stakeholders in the management of the tourist destination — it includes existing institutions, regulatory regulators, regulations and implementation of projects stimulating the participation of stakeholders.
(2) Transparency of decision-making and implementation — public access to information, process and achievements.

(3) Public–private partnerships committed to the development of the tourist destination — they consist of urbanization, digitalization, cultural programs, destination marketing, etc.

2.1.2.2 *Sustainable development*

Environmental sustainability within the development of the tourism destination

(1) Environmental impact management — water, air, soil, landscape and territories, natural resources and cultural heritage; conservation and restoration policies, mechanisms and tools.
(2) Protection of territories and resources — status, area/species, initiator and method of use.
(3) Environmental policy and development of circular and green economy — at destination and business levels — financing, investments and environmental management certificates.

Social sustainability of tourism destination development

(1) Social integration — access of socially vulnerable groups to tourist services, education, social life, organization of events and projects dedicated to cultural diversity.
(2) Preserved identity — measures for the preservation of cultural heritage, traditions and traditional events, promotion of tourist products and services related to preserved cultural heritage.
(3) Policy for stimulating creative industries and social entrepreneurship – strategies, projects, financial mechanisms and institutions.

Economic sustainability within the development of the tourist destination

(1) Contribution of tourism to the economic development of the destination — direct and indirect income, jobs and investment.
(2) Energy and resource-efficiency management — implemented measures and tools for increasing efficiency and results.
(3) Policy for the development of small and medium-sized businesses — support for innovation — including green innovations — digitalization of SMEs in tourism, support adaptation to the knowledge economy and combatting climate change.

2.1.2.3 *Hard smartness*

(1) Digitalization and modernization of infrastructure — Internet coverage and access, sensors, use of artificial intelligence and cloud services, quality of transport, communication and other infrastructure.

(2) Digitalization and application of smart technologies in tourism — AR, VR, MR, AI, QR codes, RFID, apps, platforms, before, during and after the visit to the destination.

(3) Smart technologies contributing toward sustainable development — information systems and platforms for visitors' crowd control, including redirection of tourist flows, restriction or stopping in case of exceeded norms or imposing danger to the environment and people, and application of geolocation and navigation systems.

2.1.2.4 *Soft smartness*

(1) Creativity — creative industries; business, jobs, projects, centers, clusters and "living laboratories" for innovation and creative entrepreneurship, organization of events for contemporary street art and new media, education related to culture and creative industries, and creative tourism products.

(2) Community intelligence — level of education, jobs related to innovation, smart technologies and high qualification both generally and in tourism.

(3) Innovation related to sustainable development — social entrepreneurship, green innovations in tourism and the destination, contribution of tourism to the circular economy.

2.1.2.5 *Smart ecosystem*

(1) Quality of life — quality of infrastructure, mobility, digital and transport accessibility in a destination for different social groups; digitalization of healthcare, social services and education; safety; satisfaction with tourism development; participation in the creation of tourist products.

(2) A valuable experience for tourists — satisfaction of tourists; positive image of the destination; mobility; digital and transport accessibility within the destination.

(3) Attractiveness of the destination — state and smart development/maintenance of attractions, accommodation, additional services and entertainment; accessibility to a destination.

(4) Business competitiveness — pace of development; duration of sustenance, growth, investment in innovation, technology and human resources; partner networks; joint products and platforms.

The methodology is applied in order to

(1) define the level of development of the destination as smart and sustainable;

(2) elaborate a strategy for developing the destination as smart and sustainable, based on key achievements and opportunities, as well as components that need improvement, taking into consideration the local context and trends in the global environment.

The levels of development of the destination as smart and sustainable are determined by the degree of coverage of indicators and indicators, which are divided into three groups — baseline, value-adding and exclusivity (see Tables B.1–B.3):

(1) Basic level — coverage of 100% of the basic indicators, and corresponds to a value of one (1).

(2) Medium level — coverage of basic indicators and at least 80% of those adding value, and corresponds to a value of two (2).

(3) High level — coverage of indicators and value-adding indicators and at least 70% of those for exclusivity, and corresponds to a value of three (3).

For the purposes of the study, the authors assume that all metrics and indicators have equal weight. For each of the five areas of the SSTD, the following values are determined for each of the three indicators — for achievements at the first level, a value of 1 is given, for achievements at the second level, a value of 2 is given, and for results corresponding to the third level, a value of 3 is given. These values are used to find the average score for each area. Where more than half of the indicators are non-compliant for each indicator, intermediate values may be given, for example, in the range of 0–1, 1–2 and 2–3.

2.1.3 Description of the object of study: The characteristics of the tourist destination Varna[1]

2.1.3.1 *Location and territorial scope*

The Municipality of Varna is centered around the city of Varna, which is the third largest city in Bulgaria, the largest in northern Bulgaria, and on the coast of the Black Sea. The population of Varna, according to data provided by NSI, as of December 31, 2020, is 341,516 people.

The municipality is located on the Black Sea coast, on the northern and western shores of Varna Bay, near Lake Varna. To the east, the municipality borders the Black Sea, to the north, the municipality of Aksakovo, to the west, the municipalities of Aksakovo and Beloslav, and to the south, the municipality of Avren. The city of Varna consists of five separate urban areas: Odessos, Primorski, Mladost, Vladislav Varnenchik and Asparuhovo, and five villages: Topoli, Kamenar, Kazashko, Zvezditsa and Konstantinovo.

The total territory of the Municipality of Varna is 237.5 m², which represents just over 6% of the territory of Varna district.

The city of Varna has transport, logistical, administrative, cultural, academic and economic importance for both the region and the national and international context. It is designated a functional urban area (FUA), according to the classification of European cities.

The following highways and main roads pass through the city of Varna: A-2 highway "Hemus" (E 8) from the west, A-5 highway "Black Sea" (E 3) from the south, and road I-9 (E 14) from the north. Varna Airport is an important transport hub. The southernmost and central parts of the city are connected by the Asparuhov Bridge, and south of the strait connecting the bay and the lake are the districts of Asparuhovo and Galata. An indisputable advantage from a geospatial point of view is the outlet of the Black Sea through the Bay of Varna, formed between the capes St. George and Galata.

---

[1] The data are derived from the Plan for Integrated Development of the Municipality of Varna 2021–2027.

2.1.3.2 *Connectivity with neighboring territories*

The location of the city of Varna on the coast of the Black Sea gives it the characteristic of a gateway city and a major logistics center of national and international importance. The Pan-European Corridor VIII — Adriatic Sea–Skopje–Varna–Black Sea — also passes through the destination. In addition, Varna is an important crossroad for the connection with Pan-European Corridor VII. It provides a connection between the Danube River with the Black Sea by rail and road via the line Varna–Ruse.

2.1.3.3 *Natural resources*

2.1.3.3.1 *Relief*

The main part of the territory of the Municipality of Varna is located on the Danube Plain. Its southwestern part falls within the borders of the Forebalkan. The central part of the territory is occupied by a vast lowland formed between the Frangen and Avren plateaus, on both sides of Lake Varna and the sea-lake canal. Lake Varna and the created channel lake Sea Bay divide the municipality into northern and southern parts.

2.1.3.3.2 *Climate*

The Municipality of Varna falls within the continental-Mediterranean climatic sub-region of the Danube Plain. The climate is continental, with a pronounced Mediterranean influence. The area has a unique climate. It is softened by the proximity and influence of the water basins — the Black Sea, Lake Varna and Lake Beloslav. The winter is relatively mild, the summer is cool and the autumn is long and warm. In the region of Varna, on average there are about 2,300 hours of sunshine per year, with 333 hours in July and 74 hours in December.

2.1.3.3.3 *Waters*

The most significant natural-water reservoir in the territory is Lake Varna, which is a coastal estuarine lake of natural origin. It is connected to Varna Bay and Lake Beloslav through human-made canals. It has a clean fresh surface and underground water that is used to supply the general population. There is sufficient water for production activities in the

Varna–Beloslav–Devnya agglomeration and for the development of irrigated agriculture sites.

Mineral waters in the territory are a significant resource of recreational, balneological and thermal nature. There are two aquifers of economic importance. There are 48 valid mineral-water withdrawal permits within the regions of Varna and St. Constantine and Elena and Golden Sands resorts.

2.1.3.3.4 *Soil*

The territory of the Municipality of Varna is poor in natural resources — only 6% of its total surface area is used for mineral extraction. The medicinal mud is important for the economic development of the region. Its quantities in Lake Varna are around 2,000,000 m^3.

2.1.3.3.5 *Forest resources*

The forest ecosystems in the region have an important economical, ecological and social role because they support rich flora and fauna and have key water-regulating, anti-erosion, recreational and aesthetic functions. Forests occupy nearly 30% of the territory of the Municipality of Varna, or a total of 70,357 decares. Forests and forest lands cover the Momino Coast, the Frangen plateau and the southeastern parts of the Dobrudzha plateau. The natural vegetation is represented by plantations of field ash, black alder, white willow, white poplar, hornbeam, oak, acer, acacia and hornbeam.

2.1.3.4 *Anthropogenic resources*

In the Municipality of Varna, there are material and cultural traces from all historical eras of human development. Varna falls within the territory of the Black Sea cultural and historical space which consists of Prehistory, Antiquity, Late Bulgarian Middle Ages, Byzantium, Europe and the Ottoman Empire, traditional architecture, traditions and customs, Neobaroque, Neoclassicism, Secession, Romanticism, twentieth-century internationalism and modernism, natural and cult phenomena, and underwater archaeology — sunken settlements.

Traces of early agricultural societies in Europe from the Neolithic era and ancient Odessos–Varna settlements in the Golden Sands Nature Park, architectural sites and complexes have been discovered in the region.

The Ancient City of Odessos–Varna archaeological reserve is classified as a landmark with "national importance." Its territory includes the largest Roman Baths on the Balkan Peninsula from the end of the second century AD, the Episcopal Early Christian Basilica on Khan Krum Street, the so-called "Small Baths" dating from the end of the third century AD, and the remains of fortress walls, streets, canals and public buildings.

A total of 79 archaeological sites have been registered in the territory of the city outside the scope of the Ancient City of Odessos–Varna archaeological reserve. The most significant ones are as follows:

- Varna Chalcolithic Necropolis near Lake Varna, which is one of the most significant archaeological prehistoric sites in Bulgaria, on whose territory the oldest processed gold in the world was found. It can be seen in the Archaeological Museum Varna. The found golden objects are some of the most significant relics not only for Varna and Bulgaria but also for Europe.
- The archaeological immovable cultural property Medieval Monastery in Karaach Teke area near Varna with a total area of one hectare.
- An early Christian monastery in the Dzhanavara area near Varna. It is one of the largest and most representative monuments of early Byzantine architecture on the Western Black Sea coast so far discovered.
- The Aladzha Monastery complex — a medieval rock monastery dating to the twelfth to thirteenth centuries.
- The Park-Museum of Military Friendship, 1444 "Vladislav Varnenchik," which is the location of the death of King Vladislav III Yagelo in 1444.

2.1.3.5 *Examples of park art and landscape architecture*

- Historical core of the sea garden in Varna is a wonderful representation of garden and park art. In 2018, the single immovable cultural property Black Sea Biological Station with Aquarium in Varna was established.
- Euxinograd Palace and its adjacent park were declared a group cultural property.
- The bed of Varna Bay is rich in archaeological sites, including the area of the sea in front of Cape Galata and the ancient port located over 180 decares in the Quarantineta area.

2.1.3.6   *Cultural corridors*

Varna is a part of one of the cultural corridors of Southeastern Europe — Via Pontica. It stretches along the western and southern coasts of the Black Sea and passes through Turkey, Bulgaria and Romania. In this cultural corridor, scientists have observed and analyzed the largest historical movement and layering of cultures. Via Pontica is also the second largest European migration route for birds that nest in northeastern Europe and fly south during the winter season.

2.1.3.7   *Economic profile of the municipality of Varna*

Varna is the largest economic center in northern Bulgaria. The district is strongly dominated by the diverse economy of the city of Varna and supported by the industry in its periphery.

There are three specific features of Varna which shape the dynamics and potential of the local economy. The first is the development of tourism. At least 820,000 foreigners visit the Municipality of Varna a year, mainly during the summer season. The second is the strong position of higher education in Varna and the constant flow of young people, who have contributed to the size of the "higher" economy, exceeding 225 million euros. The intertwining of science and business is the third key factor shaping the city's potential. In the fields of medicine, information technology, transport and shipping, the activities of local businesses and investment decisions are met with increased interest among young people and foreign students (Institute for Market Economics, 2020).

The economy of Varna is heterogeneous, as its structure is strongly influenced by the sea coast, visible in the relatively large role of tourism as well as the specific features in the maritime economy, trade, transport and construction. The largest sector in Varna is commerce with 20.6% of added value. In second and third place are construction (11.3% of added value) and industry (11.1% of added value), followed by transport (8.3% of added value) and hotels and restaurants (8.2% of added value).

2.1.3.8   *Description of the survey*

Based on the key parameters of a smart tourism destination (STD) outlined in the first chapter, the proposed model for a SSTD, and the SSTD evaluation criteria presented earlier, a comprehensive questionnaire

Table 2.1. Description of the survey among stakeholders in Varna.

Survey	Target group
Survey of the opinion of experts regarding the development of Varna as a STD	Experts in the field of tourism — tour operators, hoteliers, restaurateurs and tour guides, as well as experts from educational institutions in Varna
Survey of tourists' opinion regarding the development of Varna as a STD	English, German, Russian, French, Bulgarian and other tourists who have visited the destination Varna, aged over 18
Survey of citizens' opinions regarding the development of Varna as a STD	Citizens residing in the territory of Varna — representatives of different sexes, nationality, place of residence, aged over eighteen

survey has been developed. It is divided into three parts and includes the point of view of experts, tourists and citizens of Varna (see Table 2.1). The main objective of the study is to determine the level of development of Varna as an SSTD based on the opinions of the representatives of the three main stakeholders.

The main hypotheses are as follows:

H3: The city of Varna has the potential for development as a STD.

H4: The city of Varna is in the initial phase of development as an intelligent and sustainable tourist destination.

H5: There are differences between the opinions of experts, locals and tourists regarding the assessment of the development of Varna as smart and sustainable.

2.1.4 Experts' study

The sample of the survey includes 38 experts in the field of tourism — tour operators, hoteliers, restaurateurs and tour guides, as well as experts from educational institutions in Varna. The respondents have many years (at least 10) of professional and practical experience and in-depth theoretical knowledge in the field of tourism. The study includes the following main elements:

- *Method of data collection*: The data were collected through two main methods — "face-to-face" and online through an email campaign.

Table 2.2. Thematic parts of the questionnaire for experts and their role in the survey.

Thematic part	Purpose
The business/sector you represent	Collection of quantitative and qualitative data on the use of various components of smart tourism in the tourism business within Varna
Destination Varna	Receiving an up-to-date assessment of destination Varna, such as SSTD and recommendations for its successful development

- *Period and duration of the study*: From August 1, 2021, to October 31, 2022 (14 months).
- *Structure of the questionnaire*: The questionnaire is divided into two thematic parts, which are (1) the business/sector that "you" represent and (2) destination Varna (see Table 2.2).

The questionnaire consists of twelve questions, distributed as follows: first part, four questions (two opened and two closed), and second part, eight questions (four closed and four opened).

In total, 60 questionnaires were distributed, of which 42 were filled in and 38 were valid. The surveys were processed and analyzed using the statistical tools Excel and SPSS.

2.1.5 Survey of the opinions of tourists

The survey aimed at tourists visiting Varna has a representative sample of 180 respondents. Participants in the survey have the following distribution by nationality: 41 German tourists, 19 Russian tourists, 5 English tourists, 24 French tourists, 51 Bulgarian tourists, 13 Romanian tourists, 2 Ukrainian tourists and 25 others (Polish, Norwegian, Belgian, Dutch, Serbian and Austrian):

- *Method of data collection*: Face-to-face, with a printed questionnaire, and online via a link to the questionnaire.
- *Survey period*: August 1, 2021, to September 1, 2022.
- *Structure of the survey*: Introductory part, assessment of the elements of SSTD in Varna, guidelines for improvement of the city as an SSTD, demographic characteristics of the respondents (see Table 2.3).

Table 2.3. Thematic parts of the questionnaire for tourists and their role in the survey.

Thematic part	Purpose
Introduction	Collection of general information about the stay of tourists in destination Varna
Evaluation of SSTD elements in Varna	Collection of quantitative data on the current state of destination Varna as an SSTD
Guidelines for improving Varna as a SSTD	Collection of data for the successful development of Varna as an SSTD
Demographic characteristics of respondents	Collection of information of a general nature for the target group

The survey consists of 21 questions, distributed as follows: first part, five questions (one open and four closed); second part, eight questions (closed); third part, two questions (open); fourth part, six questions (closed).

The questionnaire is translated into five languages: Bulgarian, English, Russian, German and French. The questionnaires are distributed in different places in the territory of Varna in order to reach the largest possible number of tourists. Participants in the survey can fill it out in the lobbies of the tourist sites or through a link on the websites of established hotels and hostels in Varna. Paper questionnaires are also left in the Varna tourist information center and a link is shared through the website www. visit.varna.bg. The study also includes online links on social networks. In all, 185 surveys were collected, of which 180 were filled in correctly.

2.1.6 Survey of citizens' opinions

The size of the representative sample is 150 people. The questionnaire is aimed at permanent residents of Varna. It was completed by 96% Bulgarian and 4% Russian, English and Ukrainian citizens who live permanently in Varna:

- *Method of data collection*: The data were collected through two methods — face-to-face and online by distributing the questionnaire on social networks.
- *Period and duration of the study*: August 1, 2021 to September 1, 2022 (12 months).

Table 2.4. Thematic parts of the questionnaire for citizens and their role in the survey.

Thematic parts	Purpose
Evaluation of SSTD components in Varna	Collection of quantitative data on the current state of Varna as an SSTD based on an assessment of each of the five main criteria of the smart destination
Guidelines for improving the city of Varna as an SSTD	Collection of quality data for the development of Varna as an SSTD
Demographic characteristics of the respondents	Collection of information of a general nature for the target group

- *Structure of the questionnaire*: The questionnaire consists of three main thematic parts (see Table 2.4) — evaluation of the elements for SSTD in the city of Varna, guidelines for improvement of the city of Varna as an SSTD, and the demographic characteristics of the respondents.

The questionnaire includes 22 questions, distributed as follows: first part, ten questions (closed); second part, five questions (one closed and four open); third part, seven questions (closed).

In all, 200 questionnaires were returned, of which 150 were filled in completely and correctly. They were subsequently processed and analyzed with Excel and SPSS.

2.2. Experts' Research: Analysis and Evaluation of Results

The experts' comments on the use of digital technologies in all nine categories are too ambiguous (see Fig. 2.3).

They give a rather negative assessment of the use of virtual reality (Q1.1), with 42% of them choosing the answer "too limited." Of all the technologies, experts have the least information (do not know) about how much augmented reality (Q1.2) is used in the tourist destination, with 50% choosing the answer "no."

Sensors (Q1.3) for motion, biometric recognition and on/off were assessed ambiguously — 36.84% of respondents answered negatively, and another 31.58% gave the answer "yes, but not enough."

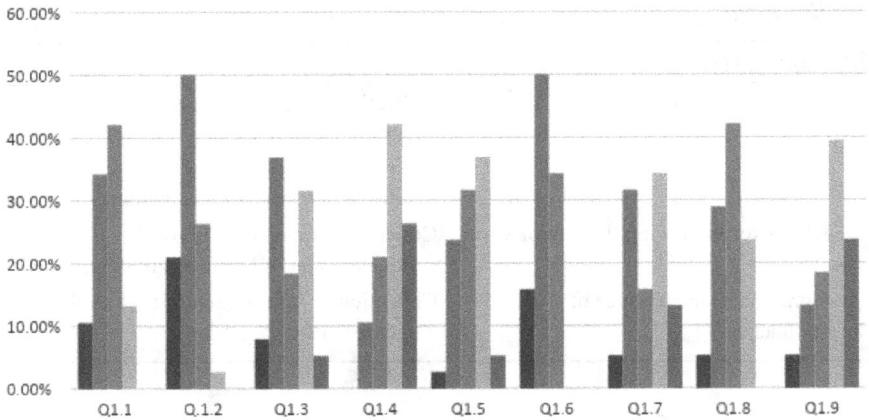

Fig. 2.3. Summary of the assessment of the use of modern smart technologies in the business of the surveyed experts.

According to experts, the most commonly utilized technologies in their industry are the wireless technologies used for (Q1.4) payment, data transmission and sending messages, registration, and navigation, including interactive maps through a web-based geographic information system. Approximately 26% indicate that these technologies are widespread in the industry, while 42% claim that their use is somewhat limited.

Respondents rated the use of translation technologies from and into a foreign language (Q1.5) as insufficient (approximately 37%).

There are mostly negative responses in the subcategory "Technologies for servicing specific groups — the blind, the deaf, those with reduced mobility, and those with special needs" (Q1.6). Half of the participants indicated that they do not use such technologies in their business.

The distribution between "yes" and "no" for "24/7 service technologies, including virtual assistants" (Q1.7) is almost equal.

According to experts, "Virtual Panoramic Tours" (Q1.8) are used sparingly (42%), but another 24% of the respondents gave an affirmative answer that there is more to be desired.

The experts considered that "Mobile applications and websites for mobile devices" (Q1.9) are not sufficiently used (40%).

Table 2.5 shows the mode and the mean value of the score for the degree (1 — low, 5 — high) in which partnership between stakeholders takes place.

Table 2.5. Summary assessment by indicators of partnership between stakeholders in tourism.

Question	Mode	Mean
Partnership between different stakeholders — business, non-governmental sector, public institutions, educational institutions, citizens	3	3.24
Partnership between different business organizations	3	3.42
Collective tourist products or created joint values/offers, including events	3	3.05
Collective mobile applications	2	2.16
Joint marketing campaigns or marketing communications, including "smart city map"	3	2.76
A system that allows information to be exchanged — quantitative and qualitative data	3	2.68
Meetings held, including virtually	4	3.47

The results show that the least appreciated according to experts is the availability of collective mobile applications. The experts consider that the elements of partnership between stakeholders are on average present in the territory of the destination. The lowest score (average value of 2.68) was given to the construction and implementation of an information system that allows the exchange of information — quantitative and qualitative data, joint marketing campaigns or marketing communications (average value of 2.76), including a "smart city map," and the possibility of publishing documents through which all interested parties can receive up-to-date information about the destination — reports, plans and strategies, financial statements, policies, and the results of meetings.

The element "meetings held," including virtual ones, receives the highest average value (3.47) and mode (4). After this, with a small difference, experts ranked the variable "partnership between different stakeholders — business, nongovernmental sector, public institutions, educational institutions, citizens."

Based on the obtained results, it can be summarized that experts in the field of tourism report that there is relatively good communication with each other, but at the same time, there is a lack of generally available and summarized information and statistics on activities in the destination.

Participants in the survey believed that elements of smart mobility are relatively well developed and implemented in the destination. The mean is 3 and the mode is 4 (see Table 2.6).

Table 2.6. Summary assessment of the elements of smart infrastructure and smart mobility in Varna.

Question	Mode	Mean
Allows control of vehicle traffic	3	2.61
Allows easy navigation in the city and the locality	4	3.29
Contributes to the fight against climate change — suitable for bicycle traffic, convenient and fast public transport for residents of and visitors to the city, with charging places for electric cars, etc.	2	2.47
Allows a high level of mobility — intermodal transport, easy movement by public transport, bicycle or on foot	4	3.16
Contributes to the prevention of environmental pollution — appropriate zoning, combination with greenbelts and areas preventing traffic jams	2	2.29
Provides access to data and communications — Wi-Fi in public places, including buses, information kiosks	3	3.05
Provides access to fast and high-quality Internet and allows the maintenance of cloud services	4	3.18

The most highly rated element is "infrastructure facilitating navigation within and around the city" (average value 3.29). The opinion of experts on the possibility of combining different modes of transportation is also positive. The control of the traffic in Varna has the lowest mean value (2.61). This is a logical consequence and a common problem in large cities in Bulgaria and requires careful planning.

Experts believe that issues related to ecology, maintaining the cleanliness of the city and creating infrastructure contributing to the fight against climate change (suitable for bicycle traffic, convenient and fast public transport for residents of and visitors to the city, with places to charge an electric car) are of utmost importance. On both questions related to sustainable development, experts have given low-score answers. The means are 2.47 and 2.29, respectively, and the mode is 2 for both. Respondents believe that the Municipality of Varna lacks infrastructure that helps fight climate change, protects against environmental pollution and keeps the city clean. Therefore, the government should pay special attention to this factor if it is to consolidate the SSTD status of the destination.

Figure 2.4 shows the respondents' answers to the question "To what extent are QR Codes used in the Municipality of Varna for better and

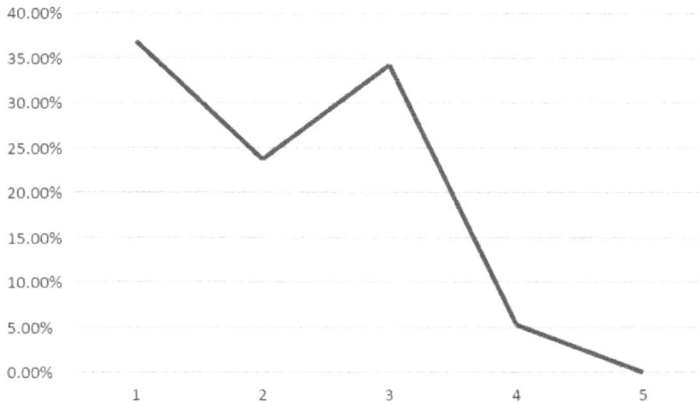

Fig. 2.4. Distribution of answers to the question "To what extent are QR Codes used in Varna for better and faster access to information about tourist attractions and places?"

faster access to information about tourist attractions and places?" where 1 represents "I do not know with accuracy" and 5 is "a lot."

The graph shows that negative responses are predominant. Only 5% of the participants indicated that QR codes are used in many places in the city.

Experts gave a variety of opinions when assessing whether the infrastructure in Varna is intelligent (see Fig. 2.5).

The accessibility, spread and speed of Internet services received good ratings. The presentation of various videos and animations about the destination is the most used smart technology in Varna by the representatives of the tourism-business sector. According to 42% of the participants, real-time bookings are used, but the application is too limited. More than half of the respondents believe that virtual tours are part of the technologies offered in Varna. At the same time, 40% do not know that there are such technologies, and 23% believe that interactive maps are not used enough. Energy management is not a widely used technology within Varna. Respondents also believe that innovations are being rarely implemented to control the overcrowding in tourist places. In general, experts are least familiar with control as well as with interactive maps for localization through a web-based geographic information system. Half of the respondents believe that no technologies are used at all to serve special

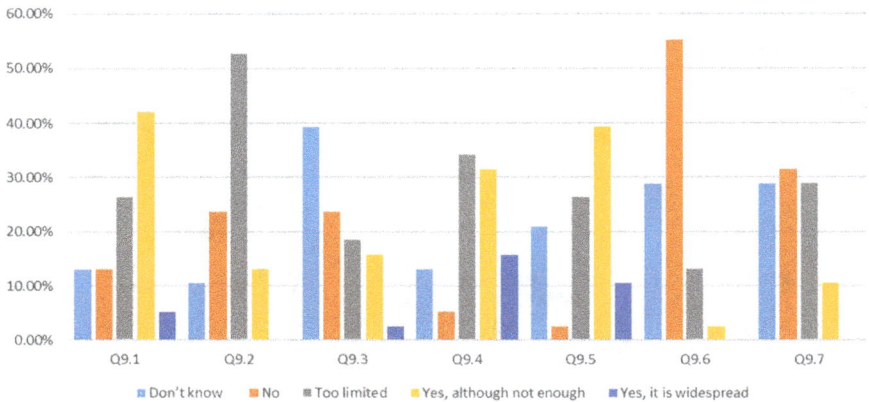

Fig. 2.5. Distribution of answers to the question "Are smart technologies used in Varna?"

groups — the blind, the deaf, those with reduced mobility and those with special needs. The participants have the least information on how much augmented reality is used in the tourist destination.

The experts assess the degree (1 — low, 5 — high) of intelligence of the management of Varna as a tourist destination, according to the indicators presented in Table 2.7.

The experts' ratings regarding the management of the destination are relatively low. The mode is dominantly 3, and the mean varies between 2.21 and 2.89. None of the answers pass the mean value. According to the respondents, Varna is doing the least in terms of e-government indicators like the online and electronic issuance of documents, services, payments and registration.

The experts give the highest scores to the implementation of a communication strategy at different units and levels. The initiation of public–private partnership by the managing authorities in the destination has the highest mean (2.97). The publication of documents through which all interested parties can access up-to-date destination data — reports, plans, strategies, etc. — has been similarly assessed with a score of 2.87.

In the answers to the open-ended questions, the experts give suggestions regarding the necessary changes for the successful development of Varna as an SSTD.

Table 2.7.   Summary assessment of the management of Varna as a tourist destination.

Indicator	Mode	Mean
Allows participation of all stakeholders in decision-making	3	2.79
Provides transparency in decision-making	3	2.45
Supports the creation of a network of business partners, public representatives, non-profit organizations and educational institutions	3	2.82
Initiates a public–private partnership	3	2.97
Provides access to information and data	3	2.89
Meets indicators for e-government — online and electronic issuance of documents, implementation of services, payments, registration	2	2.21
Valorises natural heritage and enhances ecological culture	4	2.80
Publication of documents through which all interested parties can receive up-to-date information about the destination — reports (including financial reports), plans and strategies, policies, results of meetings held	3	2.87

Most opinions and advice given are related to the element of transport, despite the relatively high assessment of the variable mobility. The need to rehabilitate the road network (streets and pavements) and create buffer parking lots is emphasized. It is imperative to increase the comfort and the cleanliness of public transport, as well as to better maintain the ticket machines. According to the participants, there is a necessity for more convenient, faster and more frequent transport to the resorts, especially to the Albena resort, as well as provision of special parking spaces for tourist buses in the city. Experts believe that a higher level of mobility is needed, facilitating and easing traffic during peak hours in order to avoid traffic jams. There is a lack of synchronization in the right to use prepaid cards for integrated public transport on urban lines operated by more than one transport company. Bicycle transport should be included in the regulation of movement outside pedestrian areas. It is also necessary for the government to ensure control over traffic and the use of bicycles and highly popular electric scooters.

The experts gave a high degree of importance to not only the availability of fast and high-quality Internet but also the ability to maintain cloud services and provide free access to Wi-Fi in all public places, including the Sea Garden, on public buses and at bus stops. The experts consider that it is imperative to improve access to information, including through information kiosks.

In terms of ranking the importance of digital technologies to be introduced and used by the tourism business, experts clearly put virtual reality and virtual panoramic tours in the first place (76%), including answers about technologies for 24/7 service and virtual assistants. In second place with the most positive responses (34%) are technologies for serving people with disabilities and other special needs. They are followed by augmented reality, mobile applications and websites, wireless technologies, simultaneous translation technologies from and into foreign languages, and QR codes. The experts put in last place sensors (for example, for motion and biometrics).

More than 65% of respondents indicated their preferences for some technologies that should be used in Varna as a tourist destination, ranking them in importance as follows: interactive maps through a web-based geographic information system for localization; efficient energy management — lighting, public-space heating and transport; technology for crowd control — events, queues and tourist places; and videos and animations about the destination. At the same time, technologies for "real-time reservation — integrated reservation systems on the destination website," "virtual tours" and "mobile applications and sites for mobile devices" remain outside the focus of the respondents' attention.

According to the respondents, museums need the integration and use of QR codes the most, both in the city (the Archaeological Museum, the Naval Museum, the Museum of New History of Varna, the Roman Baths, the Observatory, etc.) and in the vicinity (the Aladzha Monastery, Pobiti Kamani). Following this are emblematic temples, such as the Cathedral of the Assumption and the Church of St. Nicholas. Experts in tourism believe that the list of such sites should be enriched with buildings of particular architectural value, such as the Drama Theater in Varna, the Art Gallery, the remains of the Roman walls of old Odessos, marked along Tsarigradsko shose Blvd. Knyaz Boris I, Evksinograd, the building of the railway station, and the central part of the Sea Garden.

The respondents express a similar opinion regarding the need for a quality partnership between different stakeholders — business, the non-governmental sector, public institutions, educational institutions, citizens and different business organizations. In this regard, it is important for the participants to carry out campaigns with joint tourist offers and to create collective tourist products and mobile applications.

The intelligent management of Varna requires, according to the experts, the following actions:

- consolidation of the efforts of the tourism business in partnership with the municipality;
- transparency of the decision-making process and responsibility on the part of the management of the city of Varna;
- more access and public participation in the management of local tourism;
- more advertising for the sights and recreation areas that the city offers;
- wider access to information and data about the destination.

Experts outline the strong need to introduce an environmental policy related to transport and waste collection and to ban changing the appearance of valuable and significant cultural buildings.

2.3. Analysis and Evaluation of the Results of a Survey of Tourists' Opinions

The demographic distribution of the 180 surveyed tourists is as follows:

- Eighty-eight are men (48.9%) and 92 are women (51.1%).
- Seven of the respondents are 18 years old (3.9%), 15 are between 19 and 25 (8.3%), 40 are between 26 and 35 (22.2%), 57 are between 36 and 45 (31.7%), 27 are between 46 and 55 (15%), 17 are between 56 and 65 (8.9%) and 18 are 66 and over (10%).
- Thirty-four of the participants are single, 56 are in a relationship, 22 are married without children, 62 are married with children, and 6 have another marital status.
- One of the respondents has a basic education, 13 have completed secondary normal school, 23 have completed a secondary special/ vocational school, 68 hold a bachelor's degree, 62 have a master's degree, and 13 have a doctorate or higher.
- Thirteen of the tourists are still studying, 23 are retired, 3 are unemployed, 13 are administrative employees, 36 occupy a managerial position, 32 work as a freelancer, 9 are workers in manufacturing, agriculture and transport, 17 are executive staff in the service sector, 16 manage their own business, 14 are specialists in the field of education, health and culture, and 4 undertake other activities.

The results of the survey show that the respondents are a representative sample of typical individual visitors to Varna. Most are family tourists, of active age, occupying managerial positions and holding a university degree.

The introductory question of the survey refers to what tourists associate Varna with. The answers are varied as the most common words used are sun, sea, beach, rest, relaxation, vacation, entertainment and music festivals. The city is also associated with local attractions and emblematic places, such as the Sea Garden, the promenade, the summer theater, the Sports Hall, the Cathedral, the Roman Baths and the Varna Necropolis. To express emotions describing their stay in Varna, some of the tourists used such words as wonderful, beautiful, relaxing, happiness, freedom and friendly. The associations of tourists regarding the city are highly positive. Varna is perceived as a beautiful seaside destination for recreation and entertainment and to some extent for cultural activities.

The analysis of the tourist behavior shows the following distribution of responses (see Fig. 2.6).

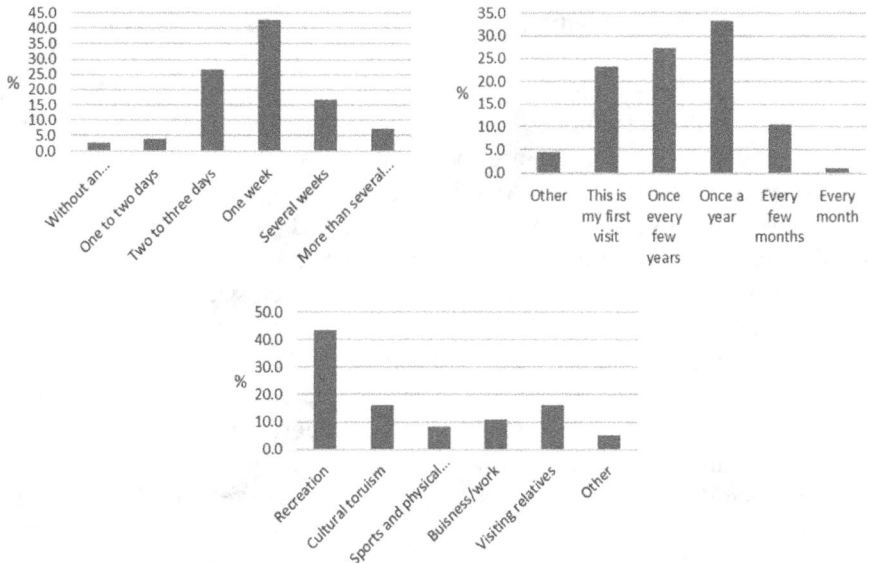

Fig. 2.6. Characteristics of the behavior of tourists in Varna: Frequency, duration and purpose of visit.

It is notable that the tourists who participated in the survey are relatively familiar with Varna. About 23% were visiting the seaside city for the first time, while 33% return every year and 27% every few years. About 10% come to Varna every few months, and approximately 1% come each month. The average stay of visitors varies from two to three days to several weeks. The majority of tourists prefer to stay for a week (33%). Respondents rarely visit the city without an overnight stay (2.8%). The main purpose of the visit is recreation (43.3%). This is followed by cultural tourism (17.1%), visiting relatives (16.1%) and business/work (10.9%). Based on the results presented in Fig. 2.6, we can conclude that the average tourist visits Varna once a year. The duration of stay is around one week and the main motive is recreation.

The visitors to Varna use different sources of information related to the destination they visit. Most often they seek the opinion and recommendations of relatives and friends. The second most important source is social media, and the third are the tour operators and travel agencies. Less-used sources are specialized tourist websites, reservation systems/websites and the website of the Municipality of Varna. Respondents rarely acquire information about the destination from the tourist information center and from paper brochures. As additional sources, some mention the website visit.varna.bg.

The second part of the survey focuses on the assessment of tourists for the determinants of SSTD in Varna. The first presented criterion is "accessibility" (see Table 2.8).

The mode for all answers is 4, and the mean is over 3.5, therefore we can conclude that respondents generally agree with the 5 main statements presented. The highest-rated element is access to accommodation and food services. Within the framework of Varna, there are many year-round

Table 2.8. Mode and mean of the elements of the "accessibility" criterion for Varna (tourists).

Indicator	Mode	Mean
Access to Internet/Internet services	5	3.87
Acquiring information in real time	4	3.59
Access to different forms of urban transport	4	3.60
Access to accommodations and food services	5	4.16
Access to natural and anthropogenic/cultural/historical landmarks	5	4.14

places for accommodation and food services, most of which have their own parking areas and long working hours. The second place, with a small difference, goes to the access to natural and anthropogenic (cultural and historical) landmarks. The tourists put in last place the receipt of real-time information, which is essential for the development of any successful smart destination.

The distribution of the answers for the variable "access of people with special needs, mothers with children, and representatives of the third age to the sights located in Varna" is presented in Fig. 2.7.

Approximately half of the tourists cannot decide whether access by people with special needs to the tourist sites is easy or difficult (mode 3, mean 3.1). A further 25% consider it somewhat easy. A relatively small percentage of respondents (15%) consider that there is limited access for people with special needs. This may be due to the lack of information or signs in the different tourist spots regarding the availability of special aid devices.

The indicator "mobility" (see Table 2.9) is analyzed through four main statements. The mean of the answers surpasses 3.5, and the mode of all questions is 4, therefore tourists perceive the presented elements some-what positively. The highest rating is given to "movement in the destination through the use of private vehicles" (mean value 3.92). Despite the lack of separate parking lots for tourist purposes, most of the sights in the territory have a suitable number of parking spaces.

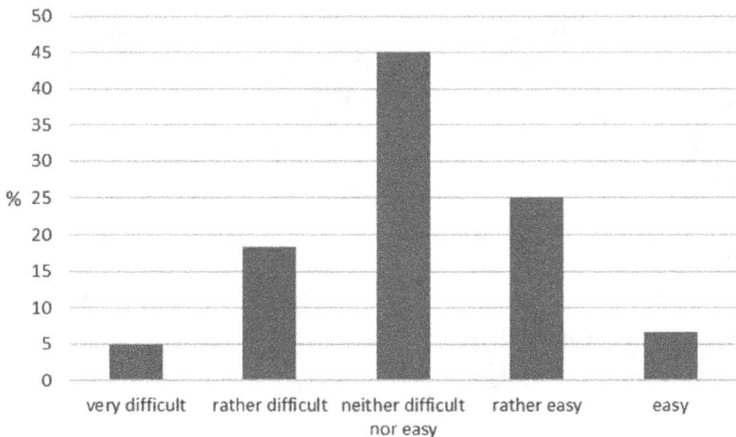

Fig. 2.7.   Assessment of accessibility to sights in Varna for people with special needs.

Table 2.9. Mode and mean value of the elements of the indicator "Mobility" for tourists in Varna.

Question	Mode	Mean
Moving within the destination using the urban transport network	4	3.80
Moving within the destination with private transport	4	3.92
Easy purchase of tickets	4	3.67
Easy-to-find information about planning a trip to the destination	4	3.77

Table 2.10. Mode and mean value of the elements of the indicator "digitalization" for Varna (tourists).

Question	Mode	Mean
Free Wi-Fi is available in Varna and there are plenty of places with free Wi-Fi access	5	3.82
Visitors to tourist attractions or places of interest can use a QR code	3	2.84
Tourists in Varna can use unique and multilingual mobile applications for solo tours in the city and the region	3	2.92
Digital and online services such as banking, shopping, rent and parking are available for tourists	3	3.07
In Varna, technologies are applied to control crowds in a given place	3	2.49
The information for tourists is adapted to the interests of the seeker, is in a preferred language and is up to date	4	3.47

The tourists rank in second place "the movement within the destination by using the urban transport network" and in third place, with a mean value of 3.77, "finding information about planning a trip in the destination." The least appreciated element of "mobility" is purchasing tickets. Tourists probably face difficulties while using the machines to buy tickets for public transport. They can be related to the inability to combine bus trips and tickets of the two competitors within Varna.

The indicator "digitalization" (see Table 2.10) is scored lower than the previous ones. It has mean values ranging from 2.49 to 3.82. Tourists agree the least with the statement that there are working crowd-control technologies in Varna. This answer is logical because, even if there is such a practice in the city, it is not promoted or known to the mass public.

Also, there is no evidence of a direct effect from it. The use of QR codes at tourist attractions or places of interest (mean 2.84) and the use of unique and multilingual mobile applications for solo tours in the city and the region (mean 2.92) were also poorly evaluated. QR codes at Varna are hardly noticeable and most are for advertising rather than information and educational purposes. Applications for solo tours in the city or area are very limited or even missing. Respondents do not rate very high the availability of digital and online services, such as banking, shopping and parking. These elements are essential for providing a better-quality tourist experience in the destination, and therefore more attention needs to be given to them.

In second place with a mean of 3.47, the respondents rank the element focusing on the ability of tourists to find up-to-date information adapted to their interests and in their preferred language. This score is probably due to the frequently updated information on social networks, as well as on the website visit.varna.bg.

The highest ranked element, with a mode of 5 and a mean of 3.82, is the element of "digitalization, connectivity to an available free Wi-Fi network and number of places with free Wi-Fi access in the city."

The fourth indicator related to the development of Varna as an SSTD is "sustainable development" (see Table 2.11). In this part of the questionnaire, the tourists rate how much they agree with four different statements. The results show a mean ranging between 2.68 and 4.24 and mode ranging from 3 to 5. Respondents agree with the assumption that there are available mobile self-service applications with environmentally friendly means of transport in Varna the least. This answer is logical because such applications are non-existent in the destination. According to tourists, public transport is not entirely environmentally friendly. There is such a

Table 2.11. Mode and mean of the elements of the indicator "Sustainable Development" for Varna (tourists).

Question	Mode	Mean
There are mobile applications for self-service with environmentally friendly modes of transport	3	2.68
There are enough green spaces, including parks and gardens	5	4.24
You can enjoy preserved natural heritage and clean beaches	4	3.75
Public transport is environmentally friendly	3	2.84

type of transport within the destination, but this fact is not sufficiently advertised to the visitors of the city. Respondents rate positively and believe that Varna has preserved its natural heritage and clean beaches. The statement that the destination has enough green spaces, including parks and gardens, has the highest mean (4.24) and mode (5). This assessment is positive and affirms the opinion that Varna has elements that help it establish itself as an SSTD.

The last criterion considered in the survey is "cultural heritage and creativity" (see Table 2.12). It covers five main statements. The mean of the answers ranges from 3.43 to 4.31, and the median from 3 to 5. This is one of the most positively rated elements of SSTD from the point of view of the visitors to Varna. The respondents are satisfied with the cultural and entertainment program, as well as with the events organized by the municipality. According to them, information related to visits to cultural attractions is easily accessible. Tourists generally agree with the statement that there are enough events to promote the cultural heritage in the destination. Respondents gave lower ratings (mean 3.43) to the use of new and interesting ways to explore the city and its sights. This creative element is important for the development of any tourist destination and needs more attention and investment.

The order of importance of the different indicators of a STD is presented in Table 2.13.

For the tourists, the key element is "cultural heritage and creativity." It is an important part for the establishment of each tourist destination.

Table 2.12. Mode and mean value of the elements of "Cultural heritage and creativity" criterion for destination Varna (tourists).

Question	Mode	Mean
Varna offers a rich cultural and entertainment program	5	4.31
Events organized by the municipality are often held in Varna	4	4.09
The city hosts events aimed at promoting cultural heritage (workshops, lectures, tours, tastings, etc.)	4	3.79
Tourists can take advantage of new and interesting ways to explore the city and its sights (VR tours, thematic tours, tours in the form of games, etc.)	3	3.43
Information related to visiting sights in destination Varna is quickly and easily accessible	5	3.96

Table 2.13.   Ranking smart destination indicators by importance (tourists).

Place	Indicator
First	Cultural heritage and creativity
Second	Sustainability
Third	Accessibility
Fourth	Mobility
Fifth	Digitalization

According to the respondents, the other elements aim to build on and contribute to the development of the territory and the quality of the tourist experience in it.

"Sustainable development" ranks in second place. The respondents attributed third and fourth places to the elements "accessibility" and "mobility." Interestingly, the least important for tourists is the indicator "digitalization." This assessment can be considered logical because this factor is essential for the development of any SSTD, but it should not be an end in itself. Rather, digitalization should be considered a method to achieve the strategic plans.

The participants in the survey also evaluated the degree of development of Varna as a STD for each of the five indicators (see Fig. 2.8).

Once more, "cultural heritage and creativity" is rated the highest, with a mean of 3.82. Approximately 31% of the respondents gave a maximum rating of 5, and 39% a rating of 4. These results correspond to the tourists' overall view of the destination presented and analyzed earlier in this chapter. In second place is the indicator "accessibility" (mean value 3.54). According to tourists, no major difficulties in visiting different tourist sites are experienced by different groups of people. Third place is assigned to the element "mobility," with a mean of 3.32 and mode of 3. Tourists have certain difficulties in moving from one place to another within the destination, especially when using public transport. Second to last in terms of mean (3.01) is the indicator "sustainable development." Tourists consider that the "digitalization" within Varna is the least developed, with the calculated mean of the answers being 2.96.

The respondents give suggestions on how to improve the image of the destination and which existing practices are most compatible with the vision of a "smart" city. The participants in the survey gave a variety of

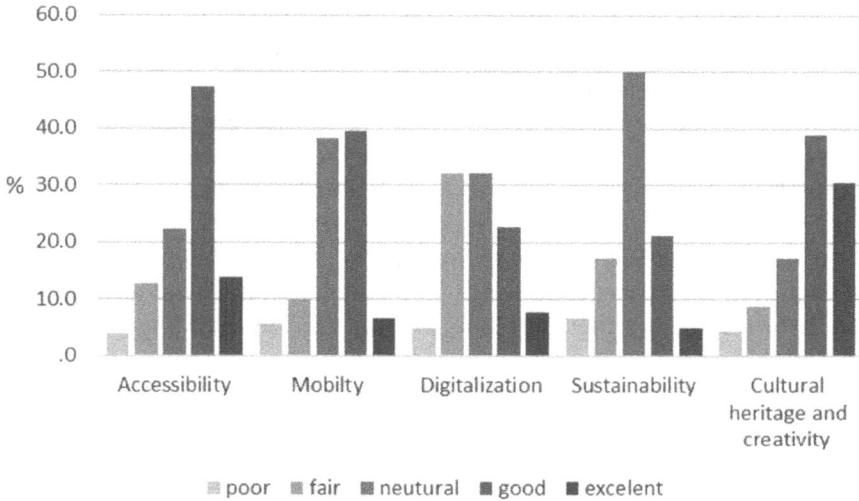

Fig. 2.8. Assessment of the level of development of Varna as an STD based on the main indicators (tourists).

different feedback. Many questioned the cleanliness of the city and addressed the poor condition of the pavement on the streets and unpaved sidewalks, especially around the bus stops. Some used negative expressions such as "the roads are in a catastrophic state and the pedestrian crossings are even worse."

A significant number of the tourists associated the existing problems with poor urban management and proposed that the local authorities take measures to increase its efficiency.

According to the respondents, improving the urban environment should include easy access to cultural attractions (including for people with special needs), more and better tourist services in different foreign languages, diversification of the cultural tourism products offered, and more information about the events in the city and good and popular hotels and restaurants. For the tourists, digitalization is also an important part of the success of STD. They believe that Varna still has a lot to do in this direction.

For the visitors to the city, it is also important that the fountains work, that there are changing cabins, toilets and showers on the beach, and that people from the service sector are more polite.

The participants in the survey consider transport to be one of the weakest elements of the destination. Varna needs better and more detailed information about public transport and more convenient connections from the airport to the city center and bus station. Tourists also add that tickets should be valid for all lines, including if they are served by different carriers. In addition, they want to be able to pay for them with a credit card, as happens in Sofia. The respondents demand transportation to the tourism sights outside the city as well as for a diversifying of the modes of transport, for instance, on water. They think that the local government has to solve the parking problem, for instance, by building multistory parking lots.

Tourists associate the Tourist Information Centre and the free walking tours offered in the city with the idea of an intelligent destination the most. Following this are the museums. Commonly mentioned are the Archaeological Museum, the Vladislav Varnenchik Museum, the Observatory and the Sea Garden. In general, the tourists also say that, for them, the idea of a smart destination is associated with transport, electric scooters, Internet access and mobile applications including parking.

2.4. Analysis and Evaluation of the Results of a Survey of Citizens' Opinions

The surveyed 150 citizens living (permanent or temporary residents) in the territory of Varna have the following demographic distribution:

- Sixty-one men and 87 women, along with 2 respondents unwilling to indicate their sex; 2 of the respondents were aged 18 and under, 43 between 19 and 25, 32 between 26 and 35, 35 between 36 and 45, 16 between 46 and 55, 17 between 56 and 65 years, and 5 were 66 and over.
- Thirty-three of the participants were single, 46 were in a relationship, 7 were married without children, 57 were married with children, and 7 had another marital status.
- One respondent had completed primary education, 12 had successfully completed secondary general education, 16 had completed secondary special/vocational school, 59 held a bachelor's degree, 48 had a master's degree, and 14 held a doctorate or higher.

- Twenty-two of the citizens in the destination were still studying, and six were retired. No one among the surveyed participants was unemployed, 11 were administrative employees, 23 occupied a managerial position, 7 worked as a freelancer, 26 were executive staff in the field of services, 28 managed their own business, 17 were specialists in the field of education, health and culture, 10 were engaged in other activities related mainly to the field of computer technology, and there were no workers in the industrial, agriculture and transport sectors.
- A hundred and forty-four (96%) of the respondents were Bulgarian citizens, two had Russian citizenship, one had English citizenship, and three had Ukrainian citizenship.

Based on the results presented above, we can conclude that the survey involved citizens with diverse demographic characteristics and profiles. This allows us to capture and analyze the opinions of a wider range of local residents.

The first presented SSTD indicator is "accessibility" (Table 2.14).

Citizens rate this indicator relatively positively, with the mean of responses ranging between 3.51 and 3.91, and the mode between 4 and 5. Participants gave the least points to the accessibility to transport. According to them, there are a number of problems in the city, such as the need for more passenger lines, the need for the construction of new parking lots, including underground ones, and the regulation of the blue zone. Accessibility to attractions and obtaining information in real time were positively evaluated. Citizens ranked first in terms of mean (3.91) the availability of the Internet, including in public places. The access of different groups to tourist sites on the territory of the destination is assessed in Table 2.15.

Table 2.14. Mode and mean of the elements of the "accessibility" indicator for Varna (citizens).

Question	Mode	Mean
Internet (including public places)	5	3.91
Information (real time)	5	3.84
Transport (public transport, taxis, bus stops and car parks)	4	3.51
Attractions in the city	5	3.87

Table 2.15.   Assessment of accessibility of tourist sites for different target groups.

Accessibility	Mode	Mean
Elderly	3	3.42
Families with kids	4	3.71
Youths	5	4.09
People with special needs	3	2.47

Table 2.16.   Mode and mean of the elements of the "mobility" indicator for Varna (citizens).

Question	Mode	Mean
Moving from one place to another	3	3.57
Easy access to public transport tickets (including via mobile apps)	5	3.51
Ease in finding transport-schedule information and trip planning	5	3.62

According to the citizens, young people and families with children have the easiest access. Older visitors face more obstacles. Respondents believe that people with special needs have the greatest difficulties related to access to tourist sites and attractions. The mean of the responses was 2.47 and the mode 3. According to the locals, this group of visitors needs specific facilities in order to be able to visit a landmark.

The second indicator considered in the survey for SSTD evaluation is "mobility" (see Table 2.16). Citizens noted three statements related to it. The mean of each of the responses is similar, ranging from 3.51 to 3.62, and the mode ranges from 3 to 5.

Out of the three statements presented, the participants in the survey rated "finding information about the transport schedule and planning a trip within the destination" the highest. This result is probably due to the introduction of various ticket-purchasing options and the partial renewal of bus stops, vehicles and public transport lines.

The following indicator rates the digitalization processes within the destination (see Table 2.17). The respondents rate it lower than the previous ones. The mean varies from 2.47 (rather disagree) to 3.06, and the mode for each answer is 3.

Users can't decide if there are enough places with free Wi-Fi in the destination.

Table 2.17. Mode and mean of the elements of the indicator "digitalization" for Varna (citizens).

Question	Mode	Mean
Free Wi-Fi is available in Varna and there are plenty of hotspots with free access	3	3.06
Visitors of tourist attractions can use QR Codes	3	2.97
There are crowd-control technologies in Varna	3	2.47

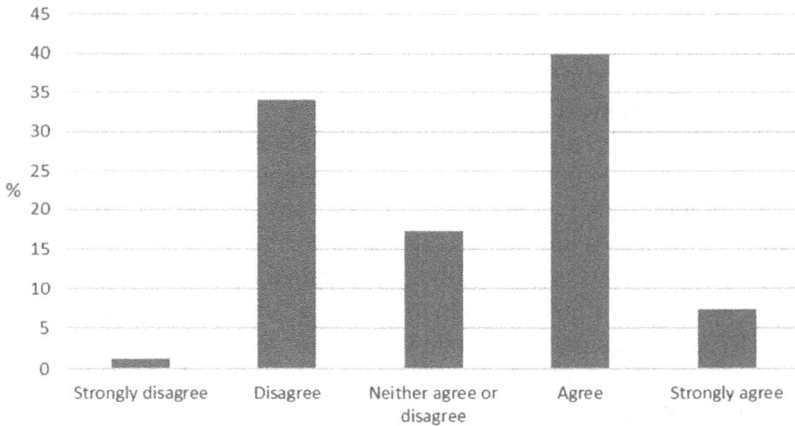

Fig. 2.9. Distribution of citizens' responses related to electronic administrative services.

According to the surveyed participants, visitors to various tourist attractions can't use QR codes. The citizens agree with the statement that there are crowd-control technologies in Varna the least. This assessment is logical because the respondents are not aware of the effects of such a technology, and such technology is most likely not actively used at all.

The electronic administrative services provided by Varna are also part of the presented indicator. The distribution of the answers to this statement is presented in Fig. 2.9.

The results are contradictory — 60 citizens (40%) somewhat agree with the statement that electronic administrative services are provided, while 34% somewhat disagree and 17.3% don't know.

The fourth STD indicator "sustainable development" is presented in the survey with three statements (see Table 2.18). Residents somewhat

Table 2.18. Mode and mean value of the elements of the indicator "sustainable development" for destination Varna (citizens).

Question	Mode	Mean
There are enough green spaces in Varna, including parks and gardens	3	3.63
In Varna, you can enjoy preserved natural heritage and clean beaches	3	2.85
Public transport is environmentally friendly	1	2.42

agree (mean 3.63) that there are enough green spaces in the city. Respondents give lower scores to the statement that the destination has preserved natural heritage and clean beaches. The citizens rank the lowest on the statement that public transport is environmentally friendly (mean value 2.42 and mode 1). Approximately 31% of respondents chose the answer "completely disagree" and 22% "rather disagree." These pronounced negative results must be taken into account by the local authority. They are evidence that, according to the citizens living in the destination, not enough measures are being taken to allow urban transport to be perceived as environmentally friendly and sustainable.

The last indicator examined in the study is "cultural heritage and creativity." Respondents rate on a scale from "bad" to "excellent" the cultural program in the city, including opera, theater, ballet and more. The distribution of responses is presented in Fig. 2.10.

None of the participants rated the program as bad, while at the same time, 40% categorized it as excellent. The mode for the question is 5 and the mean is 4.07, therefore it can be concluded that citizens consider the cultural program in the destination to be appropriate/relevant and diverse. Like tourists, residents rated the fifth indicator very highly.

In the next part of the survey, the respondents sorted in order of significance the SSTD indicators. The ranking is presented in Table 2.19.

Citizens put the indicator of "accessibility" in the first place. This is logical because the inhabitants of a city engage in economic and social activities and want quick and easy access to various public buildings, attractions, etc. The indicator "mobility" is placed second, taking into account that for every citizen the ability to quickly move from one place to another within the city, the use of intermodal transport, and convenient parking are essential.

Next in order is "sustainable development." This is directly related to the other indicators and is a guiding principle for building a modern

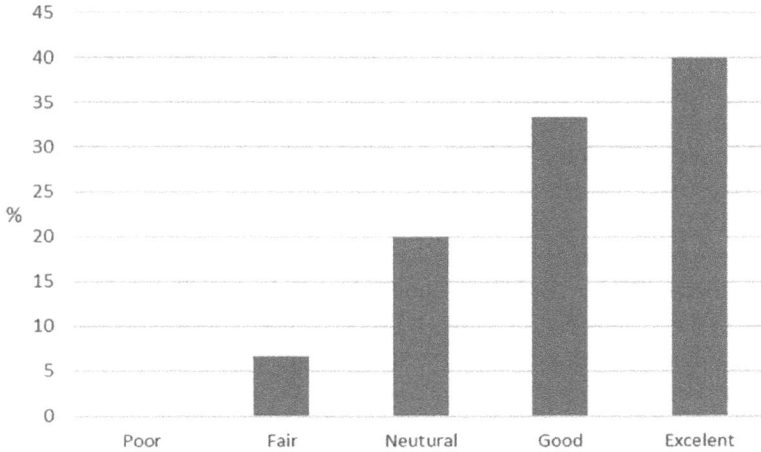

Fig. 2.10. Distribution of answers for the indicator "Cultural heritage and creativity" (citizens).

Table 2.19. Ranking by relevance of smart destination indicators (citizens).

Place	Indicator
First	Accessibility
Second	Mobility
Third	Sustainability
Fourth	Digitalization
Fifth	Cultural heritage and creativity

SSTD. The respondents rank fourth on the indicator "digitalization" as its role, as we have already mentioned, is rather supportive.

Citizens considered that the least important indicator for the development of the SSTD is "cultural heritage and creativity." Respondents probably could not always appreciate the cultural calendar/life because they take it for granted.

The questionnaire for the locals also contained questions aimed at collecting information on how to improve the development of Varna and turn it into a more successful SSTD. It is difficult for most citizens to associate the city of Varna with just one word. In some responses, we can find words expressing emotions such as joy and excitement, which shows the positive attitude of citizens toward the city.

Most people associated Varna with the sea (66.5%). In addition to the word "sea," phrases such as sea and beach, sea capital, Sea Garden, sea and tourism, sea tourism, sea and hotels are used. This clearly shows that the image of the city of Varna is mostly associated with its location. Some also associated the city with tourism (7%). The citizens used keywords, such as tourism, resort, resort town and hotels.

In all, 13.5% of the respondents enhanced the idea of the city with phrases such as "beautiful city," "dream," "young people," "hometown," "home," "freedom," "successful start," "development," "celebration," "love," "dream," "city with potential," "entertainment" and "possibilities." These answers reveal the things that the citizens search for in the destination.

The share of responses related to history and culture is relatively small (6.5%). Citizens associated the city with a "rich cultural, historical and developed destination, with the Cathedral, Varna Summer, the Dolphinarium, with museums, cultural life and sports."

The participants mentioned some negative aspects of Varna and shared their opinions. As a result, 6.5% of citizens defined the city as "polluted, noisy, expensive, boring, disorderly, overcrowded" and associated it with "traffic and chaos."

The open-ended questions enhanced those of a closed type and helped create a more in-depth analysis of the smart destination.

2.4.1 Comparison of tourists and citizens

The opinions and views of the permanent residents of the destination and its visitors are essential for its development and should be subjected to further analysis. Some of the opinions of the two groups of respondents are in opposition, while others are complementary. This allows us to build a better picture of the city and on this basis bring out a comprehensive vision for its development.

The opinions of citizens and tourists on the main criteria for SSTD were obtained on the basis of scale-type questions. By conducting a T-test for independent samples, the mean values of the answers of the two groups of respondents were compared to check whether the results were similar. There are fourteen main questions grouped into five categories. All of them have a similar null hypothesis: "Tourists and citizens have the same opinion towards/about. …"

The sample size is $N = 330$, and, respectively, citizens $n = 180$, tourists $n = 150$ and $\alpha = 0.05$.

The analysis of the answers is done with a Levin test in which H0 states that the variances are equal, and the T-test in which H0 considers the mean values of both groups to be equal (see Table 2.20).

Each criterion was presented separately, and it was analyzed whether the mean values of the responses of the two groups were significantly similar. In eight of the cases, the null hypothesis was accepted and in six was rejected.

The results of the tests show that, for eight of the questions, the opinion of citizens and tourists is similar, and for six, it differs (see Table 2.21). The similarities of the answers are mostly related to the first three indicators for SSTD, namely accessibility, mobility and digitalization. Respondents from the two groups do not have the same opinion on the indicators "sustainable development" and "cultural heritage and creativity."

The information from the survey is also enhanced by a comparative analysis of the degree of significance of the answers for each of the five indicators for a STD. For this purpose, a chi-square test was conducted. The main hypothesis H0 is that there is no difference in the ranking of the SSTD indicators by tourists and citizens. Again, we have the sample size $N = 330$, respectively, with the values: citizens $n = 180$, tourists $n = 150$ and $\alpha = 0.05$:

- *First place*: $X2(4) = 20.950$, $p < 0.001$, reject H0 (there is insufficient evidence of a relationship between the type of respondents and the order of importance).
- *Second place*: $X2(4) = 14.100$, $p = 0.007$, $p < 0.05$, reject H0 (there is insufficient evidence of a relationship between the type of respondents and the order of importance).
- *Third place*: $X2(4) = 38.347$, $p < 0.001$, reject H0 (there is insufficient evidence of a relationship between the type of respondents and the order of importance).
- *Fourth place*: $X2(4) = 15.954$, $p = 0.003$, $p < 0.05$, reject H0 (there is insufficient evidence of a relationship between the type of respondents and the order of importance).
- *Fifth place*: $X2(4) = 3.049$, $p = 0.550$, $p > 0.05$, we accept H0 (there is sufficient evidence of a relationship between the type of respondents and the order).

Table 2.20. Results of the comparative analysis of means for tourists and citizens (fourteen SSTD criteria).

Criteria	Levine test	T-test	p
Internet access	$0.985 > 0.05$ accept H0	$t(328) = -0.318$	$p = 0.750$ ($p > 0.05$, accept H0)
Access to real-time information	$0.007 < 0.05$ reject H0	$t(286.559) = -2.125$	$p = 0.034$ ($p < 0.05$, reject H0)
Access to public transport	$p < 0.001$ reject H0	$t(270.2660) = -2.125$	$p = 0.489$ ($p > 0.05$, accept H0)
Moving from one place to another — *similarity*	$0.02 < 0.05$, reject H0	$t(296.091) = 1.837$	$p = 0.67$ ($p > 0.05$, accept H0)
Easy access to tickets — *similarity*	$0.015 < 0.05$, reject H0	$t(294.927) = 1.163$	$p = 0.246$ ($p > 0.05$, accept H0)
Easy-to-find transport schedule and city-trip planning information — *similarity*	$0.001 < 0.05$, reject H0	$t(298.866) = 1.109$	$p = 0.268$ ($p > 0.05$, accept H0)
Using QR Codes — *similarity*	$0.243 > 0.05$, accept H0	$t(328) = -1.181$	$p = 0.238$ ($p > 0.05$, accept H0)
Crowd-control technologies — *similarities*	$0.035 < 0.05$, reject H0	$t(297.914) = 0.126$	$p = 0.90$ ($p > 0.05$, accept H0)
Mobile applications for self-service with environmentally friendly modes of transport — *similarity*	$0.008 < 0.05$, reject H0	$t(292.542) = 0.141$	$p = 0.888$ ($p > 0.05$, accept H0)
Free Wi-Fi network — *divergence*	$0.168 > 0.05$, accept H0	$t(328) = 5.571$	$p < 0.001$ (reject H0)
Availability of enough green spaces — *divergence*	$0.001 < 0.05$, reject H0	$t(277.956) = 5.045$	$p < 0.001$ (reject H0)
Preserved natural heritage and clean beaches — *divergence*	$0.011 < 0.05$, reject H0	$t(294.7350) = 7.385$	$p < 0.001$ (reject H0)
Ecological public transport — *divergence*	$0.00 < 0.05$ reject H0	$t(297.139) = 3.381$	$p = 0.001$ ($p < 0.05$, reject H0)
Existence of a cultural program — *divergence*	$0.498 > 0.005$ accept H0	$t(328) = -2.796$	$p = 0.05$ ($p = 0.05$, reject H0)

Table 2.21. Similarity of the mean values for citizens and tourists for the different criteria of SSTD.

Similar mean values	Non-similar mean values
• Internet access (good)	• Access to real-time information
• Access to public transport (somewhat good)	• Free Wi-Fi network
• Moving from one place to another (somewhat good)	• The presence of sufficient green space
• Easy access to tickets (somewhat good)	• Preserved natural heritage and clean beaches
• Easy-to-find transport schedule and trip planning information around the city (somewhat good)	• Ecological public transport
• Using QR codes (somewhat absent)	• Existence of a cultural program
• Crowd control technologies (somewhat absent)	
• Mobile applications for self-service with environmentally friendly modes of transport (somewhat absent)	

The results show that tourists and citizens have different opinions about the ranking of the SSTD indicators from first to fourth place. Residents of the city consider that the most important indicator is accessibility, and for visitors, this is the cultural heritage and creativity. These results highlight the differences in opinions on the development of Varna as a tourist destination and as a place to live, work and relax. Citizens demand more practicality and functionality in the destination, and tourists a more fulfilling tourist experience. This conflict must be resolved through the active intervention of both the Municipality of Varna and the tourist business. The main goal is to achieve a high degree of satisfaction for both locals and tourists. The indicator ranked last is "digitalization" — approximately 30.9% of all respondents chose to put this answer in fifth place. This result indicates that digitalization is needed for the development of the remaining SSTD indicators.

Chapter 3

Guidelines for the Development of Varna as a Smart Tourism Destination 2030

3.1. Analysis of Successfully Developed Smart Tourism Destinations

The developed methodology for the evaluation of smart and sustainable tourism destinations (SSTDs) has been applied for the analysis of two tourist destinations: Valencia and Nice. Their selection was made among European destinations that have awards and achievements related to a smart tourism destination (STD) and a smart city and are similar to Varna as a destination. In this aspect, the selection criteria are divided into two main groups:

(1) *Location*: This includes coastal destinations offering opportunities for beach holiday tourism.
(2) *Problems in and challenges to the sustainable development of tourism*: These include strongly expressed summer season, intensive traffic and destination load with visitors and vehicles, and preservation of traditions and cultural heritage.

Initially, several distinguished destinations and cities were selected, such as the smart ones — Bordeaux, Copenhagen, Valencia, Nice, Lyon and Zaragoza — which covered the exclusivity indicators used to assess SSTDs. After the application of the indicated selection criteria, Valencia and Nice were examined.

3.1.1   Tourist destination Valencia, capital of the autonomous community of Valencia, Spain

Valencia is a successful STD that cares about the environment, its residents and visitors, traditions and heritage. The city won the title of European Capital of Smart Tourism in 2022.

Valencia is the third largest city in Spain with a population of 796,549 inhabitants, located at the mouth of the Turia River in the Mediterranean Sea. The GDP per capita for 2021 is $26,254 (Spanish Statistical Institute, 2023).

One of the main historical monuments in Valencia — the Silk Exchange La Lonja — has been a UNESCO World Heritage Site since 1996 as one of the most important and well-preserved examples of Gothic urban architecture. In 2016, one of Spain's oldest popular festivals, the Valencian Las Fallas (Festival of Art, Fire and Corrida), became part of UNESCO's Intangible Heritage of Humanity, according to the Visit Valencia Foundation, along with the Tribunal de las Aguas (Water Court).

Valencia is called the city of 100 bell towers, the homeland of paella, and is also famous for the old fishing quarter of Cabañal and the neighborhood of El Carmen, which is part of the Old Town.

3.1.1.1   *Smart governance*

Valencia claims to be the first fully integrated smart city in Spain to consolidate 45 municipal services into an open-cloud platform. The cloud is used to monitor everything that happens in the city: from traffic, streetlights and parking lots to weather. For this purpose, 350 new sensors, as well as mobility data from about 3,900 pre-installed sensors and 1,000 smart traffic lights used to measure traffic density, have been installed.

The city government receives information about what is happening in the city in real time and can cancel automatic irrigation on rainy days, turn on traffic lights to let ambulances through if necessary, turn on streetlights on a cloudy day to improve visibility and let citizens know where parking spaces are available along with many other options. All these municipal services are constantly monitored (*Future Market Magazine*, 2023) and networked through the platform, resulting in greater efficiency and new quality of service. The ultimate objectives include reducing public spending, more efficient administration and improving the services the city provides. The platform additionally provides the public with a centralized, permanently accessible resource for obtaining information about municipal services. Data related to noise pollution, waste disposal, water management, air quality and many other indicators that a large city should have under control are already recorded centrally. Around 90% of the city's official forms can be completed online and residents can handle all formalities through the online platform. The city app Valencia allows any resident to interact with the municipal offices.

The current Sustainability Working Group at the Valencia Municipal Tourism Council ensures the involvement of all stakeholders.

3.1.1.2 *Sustainability*

Valencia is the first city in the world to measure and certify the carbon and water footprint (Visit Valencia, n.d.) of its tourism activity. The city facilitates the reduction of human impact with an extensive network of bike paths and a compact historic center that tourists explore on foot. It provides a bike-sharing system called Valenbisi for both visitors and residents. The system has 2,750 bicycles distributed over 276 stations throughout the city (Suspanish, 2023).

Instead of single-use plastic bottles, visitors are offered drinking-water fountains. Residents of and visitors to the city can fill their bottles with filtered and cooled tap water in one of the 18 PULSAR fountains installed by the city.

A real green liver in the heart of the city includes the Turia Gardens as well as the protected ecosystems of the La Albufera Natural Park. Valencia Bioparc is the largest zoological park in Europe, which is unique due to the fact that the animals are not caged but separated by miniature enclosures, stones, ponds and rich vegetation. Local restaurants are supplied by local farms on La Huerta and the Mediterranean.

The port area is the first in Europe to use hydrogen in its machines to reduce the environmental impact. Valencia Marina has smart supply and mooring points that allow users to track on a smartphone app the consumption of their ship or yacht. Coastal observation posts ensure that no vessel anchors in the underwater meadows of the Posidonia plant, vital to combatting climate change, damage them.

In Valencia, sea waves are already being used to generate green energy. This is done with the new wave energy converter device installed at the port canal outlet. So, in addition to the emotion of swimming or jet or surf rides, sea waves have another valuable application (Suspanish, 2023).

For the Sustainable Development Goals, Visit Valencia is teaming up with the Fundación Aguas de Valencia to provide the tourism sector and administration with tools and to advance sustainability. At the same time, the Visit Valencia Foundation's Sustainable Tourism Working Group has been established, bringing together administration, companies and universities.

Valencia prides itself on being a tolerant and accessible city, sensitive to all its visitors. Tourist buses are equipped with Navilens technology for the visually impaired. Beaches, attractions and other activities are accessible to people with physical disabilities. In many places around the city, there are pictograms for visitors with autism or intellectual disabilities. The squares and wide pedestrian spaces are ideal meeting places where tourists can mingle with locals and enjoy the local culture and lifestyle.

3.1.1.3 *Hard smartness*

Telefónica's cloud-based IoT platform forms the basis of the smart-city solution. The system complies with Fiware, a European open-source standard on which various Internet applications can be based.

Valencia's urban transport includes sustainable hybrid city buses (Visit Valencia, n.d.), which have already reduced its carbon footprint by 22%. There is a plan to add 20 new 100% electric vehicles with zero CO_2 emissions.

Smart hiking trails operate in Valencia, making walks in the city safer. These have already been installed in Valencia Marina, a cutting-edge area between the city's commercial port and the promenade with wide spaces for walking, cycling, local cuisine offerings and participation in sports and sea activities. Smart pedestrian crossings are equipped with sensors and LEDs that light up immediately as a pedestrian approaches to inform any approaching driver.

Modern design stops have been installed throughout the city, which feature Navilens technology, as well as braille signs, to guide the visually impaired. Some also offer access to Wi-Fi and USB ports for charging mobile devices.

3.1.1.4 *Soft smartness*

Valencia is a member of the Spanish Smart Cities Network (RECI). It has also been selected by the Spanish Tourism Consortium for Innovation in Tourism (SEGITTUR) as a pilot destination and is a member of regional smart destination projects, such as the 2017 Valencian Community Smart Tourism Destination Project (González-Reverte, 2019).

There are more than 400 free, high-speed public Wi-Fi hotspots in Valencia. The network, which is part of the European Union's WiFi4EU program, offers 30 MB of speed at any point and is available in squares, gardens, beaches, museums and municipal buildings. This allows residents of and visitors to the city to share their photos of iconic places with the hashtag #VisitValencia (Visit Valencia, n.d.).

Valencia is proud of the City of Arts and Sciences, which is a large complex including a planetarium, a science museum and a dendrarium. Known for its unique architecture, the complex includes L'Hemisfèric (Eye of Wisdom), an eye-shaped building designed by Spanish architect Santiago Calatrava, and L'Oceanogràfic (Underwater City), Europe's largest seaside hub, which partners with the city's main educational institution, the University of Valencia.

3.1.1.5 *Smart ecosystem (valuable, quality tourist experience, quality of life and attractiveness of the destination)*

The Urban Pulse app (*Future Market Magazine*, 2023) in Valencia provides residents with real-time access via smartphone to information about the city. The application allows you to use services such as car sharing to travel more efficiently around the city, calculating the optimal connection between two points, including all potential means of transport and always based on timetables updated in real time. Citizens of and visitors to the city can get information about air pollution, which allows them to choose the best form of transport. They can also use the app's GPS functionality to more easily meet friends. Residents can use the app to report defective municipal facilities, bad odors, flies and other problems on the relevant website. The public thus has a means of communicating directly with the administration, helping to improve the quality of life in their city.

Valencia is accessible by air, sea and land, and the city offers intermodal transport.

Valencia has an airport that connects it to about 20 European countries that served 8.53 million passengers in 2019. In 2022, it managed to restore passenger traffic to 8 million passengers.

The port of Valencia is the largest on the West Mediterranean coast and the second busiest in Spain, serving 20% of Spanish exports (ValenciaPort, 2023).

The Valencia passenger terminal meets all the requirements and specifications of the International Ship and Port Facility Security Code (ISPS), which allows mooring and servicing of high-capacity cruise ships (130,869 cruise tourists for 2021) (Visit Valencia, n.d.), yacht and catamaran rides and regular transport to the Balearic Islands.

The city also has two main railway stations, Estación del Norte, for the local train network, and Joaquín Sorolla, offering long-distance high-speed rail services with AVE trains.

Valencia has a modern road network that makes it easily accessible from any city, both in the Iberian Peninsula and in other European countries. The Mediterráneo AP-7 motorway, which runs the entire length of the Spanish Mediterranean coast from north to south, connects to the European motorway network, the A-23 Sagunto–Somport motorway and the A-3 Madrid–València motorway.

According to Spain's National Institute of Statistics (INE), Valencia reported 3,245,973 overnights for 2021 (Visit Valencia, n.d.). The total number of beds in hotels, hostels and guest houses is 21,146.

Valencia offers tourists an interactive map of the Silk Exchange, which is a UNESCO site. The map allows good orientation about the location of the site, as well as three other options: up-to-date information about the traffic in the area (which are the busy roads), the best routes for transit through the area, as well as information about bicycle routes in the area.

Valencia is an innovative city where tourist resources and offices are fully digitized. Access to any location is provided by using a QR code without the need for paper tickets. All manuals, maps and brochures are available in digital format, allowing them to be downloaded to the mobile phone for convenience during the visit. Tourist information is available in tourist offices, on WhatsApp and via live chat, and interactive touch-screens are also available in tourist offices where you can search and download information or buy tourist maps of Valencia. Visit Valencia uses resilient servers to reduce its carbon footprint. This means that when visitors buy travel services on the website, they help practice smart and sustainable tourism (Visit Valencia, n.d.).

The most sustainable way to visit the city is by foot or bike, and there are many such guided tours in Valencia. The latest tour is by bike and boat called Easy Albufera, which takes you to the Albufera Natural Park, a freshwater lagoon with unspoiled nature.

Valencia strives to offer 100% green urban "furniture." On the beach or on the marina, the visitor can sit and relax in one of the constructions made from rice husks, recycled materials and posidonia leaves (flowering marine plants) brought in by the tide. This urban furnishing is durable and sustainable, with a design inspired by ceramic mosaics and ocean waves.

The typical Valencian cuisine is also changing. Residents of the city prepare meals responsibly, and to avoid water pollution, the cooking oil used can be disposed of in recycling containers distributed throughout the city, which are equipped with sensors for more efficient management of the removal service. Tourists staying in apartments are informed that they should also not forget to recycle the oil with which they cook.

Valencia is extremely well positioned in social networks. On the city's Facebook page, there are 261,846 likes, the official Visit Valencia, Spain website has 29,000 likes and 30,000 followers, and on the Instagram page of Visit Valencia, there are 113,000 followers.

3.1.2 Tourist destination Nice, Provence-Alpes-Côte d'Azur, France

Nice is an example of a successful smart city not only in France but also in Europe. Nice is one of the pioneers in this field because it has been working to implement STD components since 2008.

In recent decades, the city has become a European leader in information and communication technologies (ICTs), biotechnology, e-tourism, e-energy and the silver economy (European Commission's Intelligent City Challenge, n.d.). In 2011, Nice won the Smarter City Challenge sponsored by IBM (IBM Smarter Cities Challenge, 2011). The city reaffirms its status as a global smart city. A study published in March 2018 by Intel and Juniper Research placed Nice in the top 20 of international smart cities. Ranked 13th, Nice is the only French city in a ranking dominated by megacities, such as Singapore, London and New York (Métropole Nice Côte d'Azur, n.d.). In 2015, it came fourth in the same ranking.

Nice is the fifth largest city in France and is located in the southeastern part of the country in the region of Provence-Alpes-Côte d'Azur. The city is the administrative center of the department of Alpes-Maritimes. According to the latest survey of the French Statistical Institute (INSEE) from the end of 2019, the population of Nice is 343,669, and with the adjacent agglomeration, it reaches almost 1 million.

From 2021, the city has been included in the UNESCO World Heritage Convention (2021). The inscribed part covers 522 hectares and includes not only the famous Promenade des Anglais but also the beautiful neighborhoods of Cimiez and Mont-Boron, numerous villas and hotels, vegetation that is both local and exotic, as well as urban planning and architecture.

Located in the building of Nice Prémium, in the heart of l'Eco-vallée of Nice, the Smart City Innovation Centre was the only one in France in 2013 with an institution with a collaborative platform supporting innovation and economic development. Its projects are in the areas of energy saving, the use of renewable energy sources, smart transportation and parking and urban monitoring. The center brings together scientists and educators, local startup companies, leading companies in the field of smart-urban management and public organizations committed to economic development and employment (Métropole Nice Côte d'Azur, n.d.).

3.1.2.1 *Smart governance*

The municipal government of Nice puts innovation and new technologies at the service of convenience, improving the quality of life of residents.

The "Smart" office (Smart Guichet) is a local digital office established in 2018, located in a municipal building in the city center. It offers users a range of services and e-services related to a variety of activities, such as education, health, social activities, sports and leisure, and is generation-oriented. Currently, this service is integrated into the municipality's website under the heading "Nice facile" ("easy in Nice"). Two virtual assistants (chatbots), Fanny and Max, have been added to the service that can provide answers to various questions from five main categories (Chatbot, Ville de Nice, n.d.).

The smartphone app Allo Mairie Nice, which brings together several applications that have been in operation for 10 years such as Blue Service and Nice Risks, aims to facilitate procedures related to whistleblowing in the city and to trigger possible intervention by the relevant authorities as quickly as possible. Citizens through Allo Mairie Nice can signal garbage disposal, pollution, improperly parked cars, technical problems and much more. This mobile application not only allows reporting but also provides information, news and services on the territory of the destination (Ville de Nice, n.d.).

The city of Nice also offers Ma Mairie Mobile (My Mobile City Hall), a mobile service for carrying out various administrative formalities in order to facilitate residents. Civil servants travel by van to different parts of the area. The site publishes a timetable of the location of the vehicle by hours and days of the week.

The opening of Nice's Open Data Portal marks an important step in the smart city's development strategy. Locals and visitors have access to a stream of information in real time. The portal is also linked to Twitter's #OpenData hashtag, the open-data news feed for the world.

3.1.2.2 *Sustainability*

By 2023, public transport provided passengers with 75% travel with reduced CO_2 emissions, and it is committed to providing 100% green/eco travel (Lignes d'Azur, n.d.) by 2025.

Tropa Verde is a platform of the URBACT program aimed at promoting eco-responsible behavior, combining a web platform and specific incentive campaigns, for example, through vouchers or prizes that motivate citizens to reuse and recycle their waste.

The RUMBLE project aims to reduce the impact of noise emissions in the port areas of the regions of the transboundary coastal zone of the Maritime Alps, the Region of Liguria, the Region of Tuscany, Corsica and Sardinia. A general problem concerning the impact of noise pollution on the quality of life of citizens is identified.

The E-health Silver Economy project is part of the strategy of the CLIP integrated thematic plan. It has several main objectives: ensuring active and healthy aging of the elderly in rural and mountainous areas; developing new pathways for the care, detection and prevention of diseases in the elderly and promoting the development of the silver economy and e-Health sector in the area.

3.1.2.3 *Hard smartness*

The project Le Monitoring Urbain Environnemental demonstrates that it is possible to test 15–30 new urban services based on a network of sensors, meters and communication network elements to collect and process information about the surrounding and urban environment (such as air quality, water, noise, waste management, lighting or vehicle traffic).

The IRIS project involves 43 European partners, with a budget of 18 million euros. With a duration of 5 years (starting from 2017), it mainly refers to the energy management of eco regions by implementing innovative integrated solutions in the fields of energy, transport, and information technology and communications. Three "lighthouse towns" in Europe have been selected: Nice, France; Utrecht, the Netherlands and Gothenburg, Sweden. The developed solutions will be implemented in four cities called "followers" (Métropole Nice Côte d'Azur, n.d.).

3.1.2.4 *Soft smartness*

An interesting project in this area is Smart Destination, launched in 2018. It aims to promote the development of cross-border tourism in the regions of the cross-border coastal zone of Var, Alpes-Maritimes, the Liguria region, the Tuscany region and Sardinia. A major problem with the existing tourism information systems is that they do not summarize the data in the same way and do not share it with each other. The project is developing in three main directions: creating a network of companies and start-ups related to tourism in order to develop a cross-border tourism sector; the development and implementation of a smart destination digital platform allowing the sharing of tourism statistics between public partners and also access by external actors to a uniform and quality cross-border, tourist database; cross-border agreement on data management and cross-promotions in cross-border territories.

3.1.2.5 *Smart ecosystem*

Nice is accessible by sea, air and land, and the city offers intermodal transport.

Nice has a port providing a ferry connection to Corsica. Nice Airport is the third busiest airport in France, located 7 km from the city center. The destination also has a railway consisting of high-speed TGV trains and local TER lines. Also passing through the territory of the region are highway A8 and national road 7.

Public transport in Nice is provided by the Lignes d'Azur, whose buses and trams cover the city, with some lines also going to nearby medieval towns, as well as to Monaco and Cannes. There is also a special bus to the mountains and express buses to the airport. Public transport in Nice forms a dense network of connections that can be used with the same ticket.

Thanks to the provision of data from the Lignes d'Azur network on the opendata.nicecotedazur.org site, Nice City Hall allows operators to develop mobile applications aimed at residents and tourists. These apps are the result of a partnership between different operators and Lignes d'Azur and are free of charge. Among them, we can list Lignes d'Azur mobile, NFC Nice Ticket, Nice Ticket, Moovit, Boogi and Transit App (Lignes d'Azur, n.d.).

The various transport links allow easy acquaintance with the sights of the city. The tourist center tries to promote tours on foot or by bike. On the territory of the destination, there are many self-service points for renting a bicycle (vélobleu) or an electric bike (e-vélobleu) (Office du Tourisme Métropolitan Nice Côte d'Azur, n.d.).

Tourism is one of the leading sectors in the destination. According to data shared by the Office du Tourisme of Nice, the city welcomes over 5 million visitors annually. In 2019, 14.5 million passengers passed through the airport. It is also the first in the country for low-cost flights. The city is the second most important cruise port in France after Marseille and the second congress city after Paris (Office du Tourisme Métropolitan Nice Côte d'Azur, n.d.). The destination has two congress centers: Acropolis et le Palais des Expositions. The city also has a substantial hotel bed base (see Table 3.1).

In the territory, there are mainly medium- and high-category hotels, as well as 22 sites of additional hotel management with a total of 3,347 rooms, according to the French National Statistical Institute. In total, the superstructure includes about 200 accommodation establishments with an approximate capacity of 10,000 rooms.

Table 3.1. Number of hotels and rooms in the city of Nice (2020).

	Hotels	Number of rooms
Total	164	10,193
1 star	2	35
2 star	19	880
3 star	58	3,284
4 star	53	5,010
5 star	3	427
Not ranked	29	557

Source: www.insee.fr/fr/statistiques.

Nice is also well represented and positioned in social networks — on the Facebook page of the city (Ville de Nice), there are 447,847 likes and 131,000 followers, and on Instagram, there are 9,956 followers (as of March 20, 2023).

The main conclusions that can be drawn on the basis of the SSTD element representation for both cities are as follows:

(1) There is a high degree of digitalization of administrative services and infrastructure, allowing the creation and maintenance of open databases, higher quality of life and partnership between stakeholders.
(2) Local governance is the main driving force of the sustainable development of a destination, but in partnership with all stakeholders: tourism, citizens, tourists, education and the ICT sector.
(3) A valuable experience for tourists is offered based on digitalization and innovation in tourism, but interrelated with improving the quality of life and the goals of sustainable development of a tourist destination.

3.2. Assessment of Varna as an SSTD

Varna is the second largest tourism destination in Bulgaria after Burgas, taking into account the number of accommodation places (405), beds (62,201) and tourists (1,009,708 registered overnight visitors) for 2022. In the district of Burgas, there are 1,086 accommodation establishments, 141,870 beds and 1,821,141 registered overnight visitors (tourists) (NSI, 2023). In all, 18.61% of Bulgaria's bed capacity is concentrated in Varna, where more than 23.27% of the total number of overnight stays in the country are realized.

3.2.1 Smart governance

3.2.1.1 *Digitalization*

Varna Municipality applies a Secure Electronic Service System, which offers the possibility of electronically requesting and executing 107 administrative services for 16 types of declarations, applications and requests.[1] For the tourism sector, it is proposed to send documents

---

[1] The data have been provided by the Municipality of Varna at the authors' request.

electronically to an email address for services related to the categorization activities of tourist sites. The e-municipality portal is used to apply for admission to nurseries and kindergartens in the first grade; municipal procurement; public registers; payment of local taxes, fees, etc. Two plat-forms have been created for the electronic submission of projects: for the electronic submission of documents for youth projects and for the submis-sion of projects financed by the Culture Fund. Varna ranks second after Sofia in Bulgaria in the share of the population using the Internet to inter-act with public institutions: 36.1% compared to 50.3% for the capital city (RSO Varna, 2022).[2]

The Tourism Directorate in the Municipality of Varna acts as an organization for the management of tourist destinations and is responsible for categorizing tourist sites, the marketing activity of Varna as a tourist destination, and collecting statistical information and maintaining a data-base on tourism in the territory of the municipality, including a municipal register of categorized tourist sites (specially developed for the needs of Varna), and registers of tourist attractions and tourist festivals and events. The municipal register provides information on accommodation and food establishments, but cannot be used to derive statistics and trends.

3.2.1.2 *Orientation to sustainable development*

For the development of Varna, a tourism development program and a schedule for the implementation of the program for sustainable tourism development for the territory of Varna Municipality 2021–2030 have been developed. The programs were prepared by a working group that includes chairmen and members of all non-governmental organizations (NGOs) in tourism, experts from higher-educational institutions and representatives of the tourism business. The documents were adopted by a decision of the city council.

The Ecology and Environment Directorate of the Municipality of Varna implements the following programs:

- Environmental Protection Programme 2021–2027;
- Energy Efficiency Programme 2021–2027;
- Waste, emissions and noise management programs.

---

[2] The data have been provided by RSO Varna at the authors' request.

With regard to strategies and plans for smart development, Varna Municipality uses the following strategic documents:

- Integrated Development Plan of Varna Municipality 2021–2027, publicly discussed;
- updated plan for integrated development of Varna Municipality 2021–2027, publicly discussed;
- the strategy for cooperation with higher-educational institutions and for the establishment of Varna as an international university city for smart development.

3.2.1.3 *Degree of democratization*

The Tourism Directorate works with the Advisory Council on Tourism, which is a permanent collective body of representatives of the administration and tourism associations. It holds publicly available information about its decisions and meetings, in addition to the regulations on the work, number and composition of the council, in a video of a meeting before the opening of the summer season 2022. By order of the mayor, a public council on tourism issues was appointed, which includes representatives of tourism, culture, sports, the airport and others. The purpose of the body is to discuss important issues related to the development of tourism and preparations for the tourist season. In the Municipality of Varna, 13 non-profit organizations are registered, which unite about 97% of the representatives of all sectors in tourism.

All decisions and information about meetings of the Municipal Council of Varna are publicly available from the official website of the council as well as through the YouTube channel of the city council.

The open database is about the state of the air, and on the website of the Ecology and Environmental Protection Directorate, there is access to a national system for providing information to the public on ambient air quality in real time and daily bulletins on the levels of the main indicators for measuring air quality. The strategic map for environmental noise on the territory of Varna is from 2017. Summarized annual information about the quality of drinking water is available on the website of Water Supply and Sewerage — Varna Ltd., with a link from the website of the Ecology and Environment Directorate.

The web-based signal system is a public module for reporting and signal management, accessible from the website of Varna Municipality from 2019 (Information Services Varna).

3.2.1.4 *Management evaluation by tourism experts*[3]

Experts are critical of the degree of smartness of destination management. Although satisfaction with meetings and opportunities for discussion and partnership between different stakeholders is considered to be average, problematic for tourism business representatives is the absence of a platform for the exchange of quantitative and qualitative data, as well as access to up-to-date information on the destination, presented in different forms. Experts are dissatisfied with the lack of partnerships with local authorities in terms of offering joint offers, products, mobile applications and advertising.

3.2.1.5 *Assessment of local residents*

Although a variety of electronic administrative services are provided in the city of Varna, citizens are not sufficiently familiar with them or want their number to increase and the quality to improve.

3.2.1.6 *Conclusions*

- The basic indicators for SSTDs are covered: a high degree of digitalization of administrative services, developed and publicly discussed strategies and plans for sustainable development, established and institutionalized dialog between stakeholders and public administration and a network of non-profit branch organizations in tourism, culture and sports, which interact with local authorities.
- There is no single portal with access to data that affects the quality of life and sustainable development of the destination, as well as monitors the results of the implementation of the strategies and plans of Varna Municipality.
- There is no active exchange of up-to-date information between stakeholders in the destination.
- There is a low degree of digitalization of administrative services related to the development of tourism.
- There are no active public–private partnerships for the purposes of tourism development and sustainable development of Varna, for example, for the functioning of the Tourist Information Centre (TIC) and portal for the creation and use of mobile applications, or for the implementation of smart tourism and smart city projects.

---

[3] The assessments of experts, tourists and citizens are derived from the results of the survey presented in Chapter 2.

3.2.2 Sustainability

3.2.2.1 *Environmental sustainability*

In the Municipality of Varna, according to the indicators of ambient air quality (Municipality of Varna, RIEW Varna), exceedances of PM_{10} standards during the winter months are registered due to the use of solid fuel for domestic heating and high wind speeds and temperature inversions, which create conditions for retention and accumulation of atmospheric pollutants in the ground air layer. Another significant factor in air pollution is road transport and dust resuspension on the road. These are all low-emitting sources close to the surface of the earth, which have a significant impact on PM levels in the environment. The share of PM_{10} emissions from domestic heating and transport emissions is roughly equal at 43.4% and 42.8%, respectively. The measures introduced to improve ambient air quality give positive results: the construction of bypass routes, the improvement of road infrastructure in order to reduce vehicle concentrations and the removal of heavy-goods vehicles from central and residential urban areas and the introduction of a blue zone. The PM midnight average concentration for 2017 and 2021 exceeded at 26 $\mu g/m^3$ with an average of 25.7 $\mu g/m^3$, while in 2021, it was 19 $\mu g/m^3$ with an average of 21.3 $\mu g/m^3$. Although in both years the exceedances of permissible PM midnight average concentration were below the maximum limit of 35 $\mu g/m^3$ per year and the concentration was below the permitted 40 $\mu g/m^3$, high values continued to be registered in February due to heating and in June and July due to increased traffic. The target of an annual average of 21.6 $\mu g/m^3$ has been achieved, but not the target for maximum exceedances to be 51.8 $\mu g/m^3$. The maximum values of 110.98 $\mu g/m^3$ and 69.10 $\mu g/m^3$ in February 2021 and 61.84 and 50.8 $\mu g/m^3$ in June 2021 were reported.

As early as 2015, there were 15 km of bike lanes in the Municipality of Varna, and their construction was part of the big project for integrated urban transport. However, they are insufficient, intermittent and mainly in the urban area. The main bike lane of the city (from the neighborhoods of Vladislavovo, Vazrazhdane and Mladost to the central part of the city) has not yet been built. There are no bike lanes to tourist sites such as Aladzha Monastery and Stone forest (Pobiti Kamani), to the Golden Sands and St. Constantine and Helena resorts or to natural landmarks, such as Rakitnika and Pasha Dere. No bike parking is available or the existing ones are not available. In Varna, there is no system for renting a bicycle through a mobile application. Only electric scooters are available, but they belong to private companies. Parking opportunities do not contribute to greener transport traffic, as there is an approximate 42% shortage of parking spaces in the city (Varna Municipality Tourism Development Program 2021–2027, updated version).

Varna is the municipality in the country with the most wastewater treatment plants (Regional Report on the State of the Environment, 2021), of which one is in the city of Varna, one is specialized for domestic and fecal wastewater and the remaining six are located in resort complexes and coastal areas from the Golden Sands resort to Galata district. WWTP Golden Sands has been reconstructed and modernized with wastewater discharge, according to the regulatory requirements, as this process continues in order to increase the treated water (72,000 PE[4] in summer and 18,000 PE in winter) and to create an opportunity to remove phosphorus and total nitrogen. Out of a total of eight WWTPs, two (Varna and Golden Sands) have permanent mechanical and biological water treatment and one has facilities that work during the summer tourist season. Currently, the sewerage system is being completed and modernized in areas such as Izgrev, Briz, St. Nikola, Trakata, Manastirski Rid, and Vayalar, where there are tourists during the summer season due to the numerous family hotels, guest houses, villas, houses with rooms for rent, etc. Measures to reduce direct discharges of untreated wastewater are applied in these areas.

In the territory of Varna Municipality, there is an installation for the mechanical and biological treatment of municipal waste and recycling of construction and demolition waste, and we can also find a composting installation for green waste and systems for the separate collection of packaging, unfit batteries and accumulators, discarded electrical and electronic equipment, scrap tires and end-of-life shoes and textiles have been built. In the regional landfill for household waste in the village of Vaglen, serving the city of Varna, composting of biodegradable waste is carried out.

Throughout the Municipality of Varna, there is an organized scheme for the transportation of mixed household waste to an installation for mechanical biological treatment (MBT) near the village of Ezerovo. In the municipal territory, there are 642 points with containers for separate waste collection; 34 are in the resorts and 289 are in the regions of Odessos and Primorski where the main tourist flows in the city are

---

[4] Population equivalent.

concentrated. The total amount of waste for the period 2016–2020 according to RIEW-Varna was 139,353,42 tons, with the waste collected remaining close to the national average and declining steadily since 2017 when it was 207,423 tons, and in 2020, it was 121,594 tons. The target of 50% recycling of paper, glass, cardboard, metal and plastics from the total amount of waste has been achieved, but the limitation of biodegradable waste has not. There are 23 sites where residents can hand over the separately collected packaging of paper, cardboard, glass and plastics, as well as 21 sites for accumulators and batteries, but information about their location is difficult to access from the website of the Ecology and Environmental Protection Directorate of Varna Municipality only and after at least four clicks and without a map indicating the locations. A system for real-time control over the implementation of the planned tasks and the work of the operators of waste collection, transportation and waste cleaning has been built. All sites of the hotel and restaurant business, including in the resorts, are covered in a system for the separate collection of biodegradable waste and their transportation to the MBT installation in the village of Ezerovo.

In the territory of Varna Municipality in 2020, there were 20 sites with certificates of origin of energy from renewable sources and 27 sites originating from renewable energy (Integrated Development Plan of Varna Municipality, 2021–2027).

A large number of the population of Varna are exposed to excessive noise levels (Integrated Development Plan of Varna Municipality, 2021–2027): 20% per L_{day}, 21% per $L_{evening}$ and 34% above the limit values for L_{night} (L_{day}, $L_{evening}$ and L_{night} are noise indicators for different parts of the day). For the city of Varna, the measured values are much higher: 47%, reaching 52% during the day. A large number of sites are subject to enhanced noise protection, and public buildings are exposed to noise levels above the limit: 63% for L_{day}, 74% for $L_{evening}$ and 72% for L_{night}. The noise pollution is due to the high intensity of road traffic, taking into account a constant increase in the number of vehicles, especially during the summer season, poor-quality road surfaces, lack of barriers for active anti-noise measures for areas with an assumed need for priority noise protection and standards or requirements for noise reduction and insulation of those most exposed to noise facades. The measures taken to reduce noise pollution, in addition to those indicated for ambient air quality, are modernization of the regulation of road traffic in the city, replacement of the public transport fleet and maintenance of the street network.

Although almost 80% of the public transport bus fleet has been reno-vated, electric buses are not yet used and only 15% are fueled with natural gas and 12% are of EURO class 6 but run on diesel. The bicycle paths have a total length of 31,870 m (0.013 m for every 100 m² area, or 13.42 km for every 100 km² area). Under the City Changer Cargo Bike (CCCB) project — City Changer Changing the City, under the Horizon 2020 Program of the European Commission with lead partner Austrian Mobility Research Agency, six electric trucks for cargo transport have been delivered, which can be used free of charge by Varna commercial establishments (Integrated Development Plan of Varna Municipality 2021–2027).

The protected areas in Varna constitute 25867.63 acres or 0.11% of its area (RIEW Varna 2022). They cover the Golden Sands Nature Park, four protected areas and one natural landmark. A visitor infrastructure has been built in the park — there are marked and well-developed alleys and rec-reation areas, six routes, old fountains and places with a charming view of the sea. A new visitor center was established in 2003. In the protected area, Aladzha Monastery, there is a tourist infrastructure — an informa-tion center and places for food and recreation.

Ecopark University Botanical Garden — Varna was established in 1977, and since 2002, it has been functioning as the first eco-park in the country, located in an area of 360 acres. The park has very good condi-tions for recreation and outdoor games.

The number of beds in accommodation establishments in the period 2019–2022, total and per unit area, in Varna Municipality, is presented in Table 3.2.

The number of beds and places of accommodation grew in the period 2019–2022, increasing stress on the destination, which could become a source of social tension through increasing noise and traffic.

Approximately 35% of the total number of private buildings in the high-energy class B are hotels that receive tax benefits for implementing energy-efficient and renewable energy measures. There is no permanent trend of increasing the issued certificates for energy savings (Integrated Development Plan of Varna Municipality 2021–2027).

3.2.2.2 *Social sustainability*

The population of Varna Municipality in 2021 was 341,737 people, which is slightly higher than that in 2020 (341,516), and at the same time, there

Table 3.2.   Number of beds and accommodation establishments in Varna Municipality (2019–2022).

	2019	2020	2021	2022
Total beds	65,943	N/a	68,712	72,411
Total accommodation (including apartments and guest rooms)	1,362	N/a	1,860	2,698
Total accommodation places over ten beds: without apartments and guest rooms	307	293	299	303
Beds per unit area: 237.5 km²	278	N/a	289	305
Accommodation per unit area	6	N/a	8	13
Accommodation without apartments and guest rooms per unit area	1.29	1.23	1.25	1.27

Source: Data from the Tourist Analysis of Destination Varna for 2019, 2020, 2021 and 2022 of the Tourism Directorate, and the Tourism Development Program in Varna Municipality 2021–2030.

has been a slight decrease of 1.05% compared to 2019. The agglomeration of Varna is 468,614 people (NSI, RSO Varna).

Approximately 80% of the attractions and tourist sites, cultural institutions and others in the municipality of Varna are not accessible to disadvantaged people.

In the city of Varna, with the support of the Culture and Spiritual Development Directorate in the Municipality of Varna, the International Music Festival "Varna Summer" is held, which has nearly 100 years of history, first happening in 1926. Other festivals that are supported by the Culture Fund and have a long history include the International Film Festival "Love Is Folly," the International Ballet Competition (since 1974) and the World Animation Festival. The city is home to one of the largest archaeological museums in Bulgaria, part of the Regional Museum of History, uniting five museums and three sites, together with four museums outside the Varna Regional Museum of History. In the municipality, there are 181,000 stock units of movable cultural value. The number of buildings of architectural, construction, art and park art is 574, with 117 architectural monuments. Among them, there are no places of global importance, while 22 are of national importance. In total, the number of buildings of immovable cultural heritage, approximately 80%, fall under an unclear category and 160 are dilapidated (Integrated Development Plan of Varna Municipality 2021–2027).

3.2.2.3 *Economic sustainability*

The GDP for Varna Municipality in 2021 was 4,289 million euros (Bulgarian currency unit) (Integrated Development Plan of Varna Municipality 2021–2027).

The share of hotels and restaurants in the value added of non-financial enterprises in Varna Municipality is 8.2%, and together with transport, trade, construction and industry account for 60% of the total value added. The growth of this sector compared to 2014 is 58%, and it is the highest represented share at the national level of 13.5%. Professional activities and scientific research have a share of 6.4%, and information and communication technologies have a share of 6.1%, which increased by 158% in the period 2014–2019. In the ICT and hotel and restaurant sectors, there was a registered growth of employees in the period 2014–2019 by 33% and 12%, respectively, with the average salary exceeding the national salary by 22% in the second sector (Integrated Development Plan of Varna Municipality 2021–2027).

3.2.2.4 *Assessment by tourism experts*

Experts are highly critical of the level of sustainable development of a destination, especially in terms of the environmental friendliness of transport and existing infrastructure, including contributing to the fight against climate change and the cleanliness of the city. Representatives of tourism are highly dissatisfied with the availability of appropriate conditions, such as infrastructure and technologies servicing people with special needs. Experts believe that sustainability is essential for the prosperity of a tourism destination.

3.2.2.5 *Assessment of residents and tourists*

Locals rate the level of sustainable development as average, which is no different from the assessment of tourists. The exception is the satisfaction of the green spaces in the city, which are more highly appreciated by tourists. Compared to all other indicators, the citizens of Varna are highly critical of the environmental friendliness of transport and the degree of preservation of natural and cultural heritage. The Varna Summer Festival is perceived by locals as emblematic and significant for the city's culture. Sustainable development ranks third in importance for the citizens of Varna, but second for the visitors.

3.2.2.6 *Conclusions*

- There is no active dialog between stakeholders — citizens, municipal management and business — and quick and easy access to data on sustainable development. On the one hand, residents are not sufficiently aware of the problems and the real situation, unlike the representatives of the tourism business, e.g., the negative and positive characteristics of the destination on the issues of environmental protection and sustainability of the economy and society. According to the presented data, there are both relatively poor results (noise, air pollution from transport and energy efficiency) and in some aspects relatively good results (wastewater treatment, growth of the ICT sector and protected areas). On the other hand, tourists who base their assessments on achievements in other destinations in Europe or their own countries show a more significant interest in the achievements related to sustainable development and at the same time rate those in Varna relatively low.
- Problem areas in terms of sustainable development include the use of alternative and environmentally friendly transport, including by tourists; access to attractions by representatives of socially vulnerable groups; ineffective control of traffic and noise; lack of certified tourist companies for environmental management and high destination load levels with beds in accommodation places.
- The achievements in terms of sustainable development include the regulation of tourist activities and flows in protected areas, the creation of opportunities for the diversification of tourist products and activities which will reduce the strong concentration of tourists during

the summer season, the development of the ICT sector and the basis for separate waste collection and their effective management, including in tourist sites.

3.2.3 Hard smartness

Varna is the main telecommunication hub in the country, through which the connections of the Trans-European Telecommunications Network with the underwater optical cable systems in the Black Sea KAFOS and BSFOCS are carried out, which allows the transit of large communication flows from Russia, Ukraine, Georgia and Transcaucasia through the territory of Bulgaria to Greece and Cyprus and through them to the Mediterranean and North Africa. BSFOCS is a Black Sea fiber-optic cable system connecting the cities of Varna (Bulgaria), Odessa (Ukraine) and Novorossiysk (Russia). The optical system has been in operation since 2001 (Integrated Development Plan of Varna Municipality, 2021–2027).

There is Internet coverage with 30+ mbps broadband speed on 99% of the territory of Varna Municipality, with 100% on the coast, but Varna ranks 13th in the country for Internet access. The share of households using the Internet at home is 75%, compared to municipalities such as Ruse and Burgas where it is 84% (Varna Municipality, 2022).

A Center for the Management of Mass Urban Public Transport (MWTP) has been built in Varna, serving as a place for continuous event logging, which receives information from vehicles in real time, processes data, calculates their arrival time at stops and transmits information to the relevant terminal devices (stops, online and mobile applications). Thanks to this center, MWTP vehicles inform passengers about traffic in real time. A system has been built to give priority to the MWTP buses along the route of the BRT corridor. Varna is the first city in the country to introduce a fully automated ticketing system in public transport. In other cities such as Sofia and Burgas, only smart cards work, and ticket machines have not yet been introduced (Varna Municipality, 2022).

In the territory of Varna, there are 10 public places in the central pedestrian zone and nearby tourist attractions with free Wi-Fi Internet access. For the purposes of better orientation and easy access, information kiosks located at the airport, railway station, pedestrian areas and nearby cultural and administrative institutions, as well as maps in the places that attract the most interest and flow of people, including transport centers, are needed (Varna Municipality, Tourism Directorate, 2022).

The VISIT Varna website (www.visit.varna.bg) is interactive, with attractive design and the user-friendliness of information provided in five languages: Bulgarian, English, Russian, German and French. Users of VISIT Varna can be informed about accessibility and transport, accommodation and meals, sights and entertainment and current events in Varna. The website provides information for tourists with different interests and motives for visiting the city. There is data on mobile applications that are useful for travelers in Bulgaria and are created under national projects. For visitors to the city, there are interesting mobile applications for renting electric scooters. The website provides the possibility of a 360-degree virtual tour of Varna (www.visitvarna360.com), made as a recording, not in real time. The VISIT Varna website, in addition to current events, does not inform about the weather, traffic conditions, opportunities for guided tours, etc. Also missing are first-person stories, shared experiences and tours, blogs and other feedback options. Although the available parking lots are described, they are not depicted on a map, and there is no access to an application for the availability of parking spaces in real time, payments and more. The information offered on the website is not structured by the type of experience and tourism, despite the many diverse offers for leisure time and description of attractions, sports and recreation conditions. The website provides an opportunity for travel entrepreneurs to register profiles on the platform, keep them up to date and send inquiries from tourists directly to their registered emails. In the period 2020–2021, the number of visitors to the site increased by 150,000, increasing the share of repeat visitors from 14.3% to 16.5%. Bulgarian is the most used language, by over 40% of the visitors (Varna Municipality, Tourism Directorate, 2022).

Several attractions and tourist sites have QR codes that were created under the national project iLove Bulgaria. In Varna, there are no attractions and places of entertainment using AR, VR and MR technologies.

An Information Register of Varna Immovable Cultural Heritage — varnaheritage.com — has been created, which presents the cultural values of eras and architects with an interactive map. A route in the central historical part of the city is described. The platform is available in Bulgarian and is practically unknown to tourists and residents of the city (Tourism Development Program in Varna Municipality, 2027).

Three 7-min tourism films have been created in four languages: *Sea Treasures*, *The Call of Civilizations* and *Varna — Magnetic and Accessible*, which are provided free of charge for broadcasting to all hoteliers and tourist sites.

3.2.3.1 *Assessment of tourism experts*

Experts assess the infrastructure in Varna as insufficiently responsive to the modern intelligent infrastructure despite the application of a number of digital technologies. They are critical of the weak and even missing use of QR codes, VR and AR, technologies to control crowds and traffic in the city and serving people with special needs. Representatives of the tourism business believe that there are very good conditions for using Internet-based services, including access to a fast Internet connection.

3.2.3.2 *Assessment of residents and tourists*

Tourists are critical of the digitalization in the city of Varna, as few appreciated the availability of QR codes and access to multilingual mobile applications for solo tours in the city and to real-time information. At the same time, they are satisfied with the provision of adapted and up-to-date information, which, according to their interests, is also very good for free access to Wi-Fi networks. There are strong negative reviews from visitors regarding the road infrastructure and urban environment, concerning its cleanliness, maintenance, quality and convenience, despite the repairs and modernization of key pedestrian areas, parks, bus stops, etc.

Residents are also critical of the digitalization in Varna, which ranks fourth in importance for them, both in terms of access to free Internet in public places and the availability of QR codes and technologies to control crowds. The citizens of Varna are however relatively satisfied with the access to the Internet and the information in real time similar to the opinion of tourists.

3.2.3.3 *Conclusions*

- A basic infrastructure for communication and service to tourists and residents exists, but it is not fully used to provide information in real time and in the context of the location, with more opportunities for feedback and customization.
- The level of infrastructure and the degree of digitalization at this stage are not in favor of creating an unforgettable experience, together with the participation of locals, businesses and various visitors, and satisfying all parties.
- Locals do not see a direct benefit from digitalization and do not associate the quality of life with the application of digital technologies.

- The changes in terms of hard smartness are in the direction of the application of digital technologies that provide benefits to tourists, citizens and business: digitalization of information services for tourists, management of the energy used in public places, services for people with special needs and digital marketing of the destination, especially among foreign tourists.

3.2.4 Soft smartness

In 2019, the number of Research and Development (R&D) personnel employed was 2,510, which is 2% of the total number of employees, a similar share to that for Bulgaria, and the expenditures were BGN 35,838,000, which is 0.5% of GDP (versus the national average of 0.83%). The expenditures for research and development in trade, transport and tourism in the Municipality of Varna are 8% of the total for the country. In the territory of the municipality, there are three research institutes: the Institute of Fisheries Resources at the Academy of Economics, the Center for Hydro and Aerodynamics and the Institute of Oceanology at the Bulgarian Academy of Sciences, whose activity is indirectly important for the development of maritime tourism (Integrated Development Plan of Varna Municipality, 2021–2027).

Varna ranks second after Sofia in the number of newly opened companies in ICT and Fintech and has emerged as a secondary destination for outsourcing in these sectors due to the increasing difficulties in finding a workforce and the high rents of offices in the capital. Varna is an attractive center for young staff from neighboring municipalities. In the period 2014–2019, the growth of the number of small companies in ICT was about 34%, which significantly exceeds the 3.9% growth of the total number of small companies in Varna Municipality for the same period. Similar is the trend in terms of employees in the business in this sector: a 46% increase from 2014 to 2019 with a total of 3,400 employees (Integrated Development Plan of Varna Municipality, 2021–2027).

The total number of students in the Municipality of Varna in six universities is about 30,000. In five universities, there are majors related to informatics and computer technologies, with a total of 2,500 students: an increase of 750 students in 2019 compared to 2014. Although the expenses for scientific and development activities in the Municipality of Varna are low compared to the average in the country, they are the most significant in the ICT sector, with a share of 1.6% of the total for the country and with a growth of 70% in the period 2014–2019.

The municipality, together with the universities in the city, the big employers and others, is implementing a memorandum for the development of Varna as a city of knowledge (Institute for Market Economics, 2020). In this regard, projects for the construction of the Industrial Zone Aksakovo, the High-Tech Innovation District in the Malka Chaika area and a new modern mathematical–technological campus with a new building for the High School of Mathematics "Dr. Petar Beron" are forthcoming.

The share of people with higher education in Varna Municipality was 25%, but in 2019 compared to 2014, it was 6% (for the country, it was approximately 15%), and the share of people with secondary and secondary special education increased by 2%. In 2021, the share of people with post-secondary education in Bulgaria was 29.6%, and for the EU, it was 33.4% (NSI, 2022).

In the Municipality of Varna, in addition to the traditional festivals, since 2015, several innovative events have been held, such as Varna Carnival, The Longest Dinner in Asparuhovo and Days of Bulgaria — a concert dedicated to the Day of Varna with a laser show and fireworks display at the Port of Varna.

According to the European Monitoring for Creative Cities, Varna has an index of 13.8 and is 36th out of a total of 39 observed cities with a population between 250,000 and 500,000 inhabitants. The assessment was made according to three groups of indicators: cultural vitality, creative economy and environment for the development of the creative and cultural economy and activity. According to all metrics and indicators, Varna is below the average for the group with the exception of "New Jobs in the Creative Economy," with an index of 23.6 and an average of 23.2, and "Cultural Attractiveness," with an index of 27.1 and an average of 27.2. The lowest is the environment for the development of creative and cultural economy and activity, with an index of 8.5 compared to an average for the group of 27.6. For comparison, Thessaloniki has an index of 17.5 and ranks 31st in the group (Cultural and Creative Cities Monitor, 2021).

3.2.4.1 *Conclusions*

- The innovation potential of Varna is determined by the developed university network and established research institutes, but is still without any direct and significant impact on the quality of life of local residents and sustainable development.

- In Varna, there is an infrastructure (which needs to be modernized) and potential for the development of the creative and cultural industry: cultural institutions, innovative cultural events, universities attracting young people from neighboring municipalities and countries, from all over the country and the EU and others, established dialogue and active support of young people by the Municipality of Varna.

- On the territory of the municipality, there are no creative ("living laboratories") and technology centers and incubators, associations, networks or areas for intelligent development and uniting science, business and local government.

3.2.5 Smart ecosystem

3.2.5.1 *Quality of life*

The schedule of the city and public transport of Varna is already part of Moovit, the first mobile and transport application worldwide. It allows travelers to receive step-by-step directions while traveling to a landmark, street or main public transport stop and to view bus and train schedules, arrival times, service notifications and detailed routes on the map so that they know exactly how to get to anywhere in Varna.

Residents and tourists in Varna can use an information system for public transport with an online map connected to a navigation system showing the arrival times of buses at stops and their location, bus stops located near the user, bus lines on specified routes through the applications Bus Varna and Varna Traffic and the website of Public Transport Ltd. (www.varnatraffic.com). Through this Internet portal, each user can create an account in which they can not only monitor bus schedule information but also activate or deactivate their transport card, view and add to their account balance, etc. The Parking Varna and Ticket Varna applications make it possible to pay for parking a car in the Blue Zone and buying a ticket for public transport. There are 45 inner-city lines with a total length of 755 km, served by bus and trolleybus, and 14 non-urban lines with a total length of 210 km. Bus rapid transit (BRT) corridor No. 1 runs from Vladislav Varnenchik to Briz residential complex, offering fast, convenient and cost-effective urban mobility through separate lanes called corridors, which is a fast mode of operation at short intervals. All 85 new buses and 35 trolleybuses (out of the total of 127 needed) are air-conditioned and low-floor, making them comfortable for disadvantaged people and mothers with baby carriages (Integrated Development Plan of Varna Municipality 2021–2027).

In the territory of Varna Municipality, several projects have been implemented in support of disadvantaged people, creating a Center for Support of People with Disabilities and Very Severe Disabilities, a Center for Care for People with Different Forms of Dementia and a Daycare Center for People with Different Forms of Dementia and Support for Their Families, as well as social housing for disadvantaged people and shelter for the temporary accommodation of homeless and socially disadvantaged people. There are 35 social services for socially vulnerable groups: 16 for adults and 19 for young people up to 18 years of age. In order to ensure access to the institutions by citizens with difficult conditions, the following projects have been implemented: eight specialized minibuses for the free transportation of people with disabilities, specialized vehicles for people with difficult mobilities and 17 rooms for children and adults, some of which are managed by NGOs, provided by a decision of the Municipal Council for Community Services (Integrated Development Plan of Varna Municipality, 2021–2027).

The average population density in Varna district is 122.9 people/km^2, which is significantly higher than the national average, and the average for northeastern Bulgaria is 63.2 people/km^2. This is due to the extremely high density of Varna Municipality, which is 1452.2 d/km^2.

The street network includes 52% fourth class, 25% third class, 20% second class and only 3% first class. It covers the entire territory of Varna Municipality, but there is still a need to reconstruct part of it (i.e., to change the class of approximately 32.7% of streets), especially in residential neighborhoods, build more smart-pedestrian crossings (currently, there are 15 to 25 intersections in the city) and install more traffic lights (Integrated Development Plan of Varna Municipality, 2021–2027).

Varna district ranks second in terms of population in Bulgaria connected to public sewerage and wastewater treatment plants: 86.5%, compared to the national average of 64.5% for WWTP and 76.1% for sewerage (RIEW Varna).

Varna ranks second in the country in health-care services and the established base in this area. The Medical University, together with

St. Marina University Hospital, is an incubator for innovations in medicine, pharmaceuticals and dentistry.

The residents of Varna rate accessibility and mobility as good, although there are fluctuations about easy movement from one place to another. For citizens, Varna is a beautiful, coastal city with potential and opportunities for entertainment, but at the same time, it is defined as expensive, noisy, crowded and untidy, with unmanageable traffic and chaos.

3.2.5.2 *Conclusions*

- In Varna, there are good conditions for quality of life at an average level and potential for improvement.
- A significant problem is the lack of infrastructure and a flexible policy that corresponds to the needs of a densely populated city, with a number of universities and tourism businesses attracting additional temporary residents.
- Transport traffic needs optimization and better-quality control, not only in terms of increasing the mobility of residents, especially during the busy summer season, but also for the purpose of preventing pollution and noise.

3.2.6 A valuable experience

Varna is accessible by three types of public transport: land, air and rail. Currently, there is no passenger transport by sea to Varna Marine Station, but there are conditions for meeting and servicing seagoing vessels to private individuals and private businesses. There are five yacht ports in the territory of the municipality. Varna is planned to become part of the comprehensive transnational European transport network by the end of 2050, with the main barriers to integrating the city into this network being the unfinished Hemus Motorway and the unbuilt high-speed roads (Varna–Ruse and Varna–Durankulak). At the same time, the city is accessible through the constructed section of the Hemus Motorway (Sofia–Veliko Tarnovo–Varna) and the Republican road I-9 (part of the Turkey–Romania land link), the first-class road I-2/E-70/Varna–Ruse and the Black Sea highway E-87, which duplicates part of the I-9. Private bus companies serve the lines Varna–Veliko Tarnovo–Sofia (all cities in northeastern Bulgaria), Varna–Burgas and several other cities in southern Bulgaria. Varna Bus Station needs modernization and substantial reconstruction, as well as a direct connection to the other transport centers: the airport, railway station and sea station. Varna Airport, with a connection to 100 destinations in 57 countries, is connected to the Hemus Motorway and E-87 and by public transport to the city and the resorts (one bus line, however, does not stop at the bus and railway stations). Varna Railway Station is mainly a passenger terminal that serves railway connections with Sofia, Burgas, Plovdiv, Pleven and Ruse.

3.2.6.1 *Assessment of tourism experts*

Experts define mobility in the destination as being relatively good, both in terms of navigation and easy movement by public transport, bike or on foot, with opportunities to combine different modes of transport. Representatives of the tourism business are critical of the opportunities for using a bicycle for movement outside Varna to resorts and landmarks, the availability of buffer parking lots suitable for tourist buses and traffic control for cars, buses, bicycles and electric scooters.

3.2.6.2 *Assessment of tourists*

The assessments of tourists for Varna are positive — it is perceived as beautiful, but its attractiveness is associated primarily with recreation and entertainment by the sea in the summer and less with the opportunities for cultural and cognitive pursuits, sports and ecotourism. Approximately 50% of the visitors choose Varna for recreation purposes. But, the city is also identified with the key sights and the music festival.

Tourists rate access to attractions and accommodation as relatively high, although in reality, 80% of them do not have the conditions to meet and serve people with special needs. This assessment is based on the existing possibilities of reaching the sites in the central part of the city by bus, bicycle and electric scooter, as well as the available pedestrian areas. It is important to note that sights such as Stone Forest (Pobiti Kamani), Rakitnika, Pasha Dere, Aladzha Monastery, and attractions and entertainment venues such as the Karting Track, Aquapark Golden Sands, the visitor center at Golden Sands Nature Park, Galata Lighthouse, Asparuhovo Beach and Park and Karantinata fishing village are difficult to access by public transport (there are no transport lines or only one line and/or the bus timetable is not easily available). The surveyed tourists cannot assess whether there are conditions for access by people with special needs because they are not clearly marked or this is beyond the interests of the visitors. At the same time, citizens who know the attractions and the urban environment give a very low rating to the access for people with special needs to attractions and tourist sites.

Tourists highly rate the cultural and entertainment program in Varna as well as the active contribution of the municipality. Although visitors to the city are satisfied with the activities promoting cultural heritage and information about the sights, they believe that there is more to be done in terms of offering new and interesting ways to explore the city and its attractions.

Although assessments of the diversity of urban transport and mobility in general are good, in the freely expressed answers, there are criticisms of transport in the city (especially to more remote landmarks) and information about it. Tourists are also critical of the promotion of events in Varna. They are left with the feeling that, in Varna, not enough measures have been applied to comply with the rules and there is little control by the local government.

Over 50% of the surveyed tourists visit Varna more than once and are willing to stay for up to 1 week. The predominant sources of information about the destination are personal contacts — personal or through social networks — and much less often communication channels managed by the Tourism Directorate and tourism business, including social media profiles.

3.2.6.3 *Conclusions*

- The existing accessibility to the destination is very good, but needs a more efficient connection to the internal transport network and diversification in terms of water transport.
- The mobility of tourists is relatively good, but the information (mainly for foreign tourists) is still not quite sufficient, and the degree of intermodality in transport is low. Also, there is no convenient transport accessibility to significant sites and interesting places, including opportunities for an alternative method of transportation to them and outside the city.

- Varna has a rich tangible and intangible cultural heritage, which should become one of the leading motives for visiting the city all year round.
- The attractions and cultural programs in Varna are a key factor for the satisfaction of tourists, but it is necessary to diversify the approaches to presenting the values of the destination (including to meet the interests of modern generations) as well as to improve its overall atmosphere.
- Varna offers a valuable experience at a satisfactory level, but in order to increase its value, it is necessary to implement new digital technologies and implement more interactive and modern cultural and leisure activities.

3.2.7 Attractiveness of the destination[5]

In the period 2015–2019, tourist demand was characterized by an increase in overnight stays in accommodation establishments (an average rate of increase of 7.5%) and a variable change in the number of nights spent (a slight decrease of 0.6% in 2017 compared to 2016 and a decrease of 4% in 2019 compared to 2018). In 2020, the situation changed radically due to the COVID-19 pandemic and flight cancellations, measures to prevent infections, etc., and there was a decline of nearly 73% in both indicators. In 2021 and 2022, there was an increase compared to 2020 both in the number of tourists and in the nights spent. The growth of overnight stays reached 57.54%, and the growth of overnight stays in 2021 was 82% more than the previous year. In 2022, the number of tourists increased by 28.9% (809,223) and overnight stays increased by 89.77% (3,486,971) compared to 2021.

On the one hand, Varna is losing its market position in Germany and the UK, but on the other hand, the number of Romanian, Bulgarian, Polish and Czech visitors is increasing. In 2022, there was a tendency to restore tourist demand levels compared to 2019 (approximately 81%), but still not enough compared to the record of 2016 in terms of number of visitors,

---

[5] The sources for this paragraph are the NSI, Municipality of Varna — Tourism Development Program of the Municipality of Varna 2021–2030, Tourism Analysis of Destination Varna 2019–2022 and Visit Varna.

which exceeded 1 million — 1,014,759 — with overnight stays of 5,059,435.

Approximately 80% of the overnight stays are between June and September. The average decreased in the period 2015–2019 from 4.6 to 4.4, and in 2022, it was 4.3.

In the Municipality of Varna, there are a variety of accommodation establishments by category and type. Approximately 75% of the bed base is concentrated in hotels, but the share of apartments and guest rooms is 74%. Almost 50% of the hotels are of high category, and 66% of the beds are concentrated there. Its average annual employment is between 20% and 24%, and the rate of effective employment is between 40% and 44%.

Despite the variety of attractions in the Municipality of Varna, there are no cultural values of world importance, and the golden treasure from the Varna Necropolis is of national importance only and not popular enough and emblematic for the destination. In the territory of the Sea Garden of Varna, there are three key sites that are related to the sea and in particular the life and specifics of the Black Sea: the Aquarium, the Natural History Museum and the Dolphinarium. At the same time, there is a low degree of interactivity and active participation in all the three sites and no use of digital technologies.

To estimate the stress on the destination, the ratio of the number of beds per 100 people of the population is considered, and the data for the period 2019–2020 are presented in Table 3.3. There is an increasing trend in this ratio in favor of the number of beds, which is a potential source of social tension in terms of increasing number of visitors and vehicles, amount of waste, noise and congested places for relaxation, entertainment and more.

Table 3.3.   Number of beds per 100 people of the population in Varna (2019–2022).

	2019	2020	2021	2022
Beds (total number)	65,943	N/a	68,712	72,411
Population	345,151	341,516	341,737	345,154
Number of beds per 100 people of the population	19	N/a	20	21

Source: Based on the National Statistical Institute, Tourism Development Program in Varna Municipality 2030, Tourist Analysis of Destination Varna.

3.2.7.1 *Assessment of tourists*

Tourists rate "cultural heritage and creativity" as the most significant for their experience and give a very good rating on this indicator for Varna.

3.2.7.2 *Conclusions*

- Varna is attractive for tourists, mainly for recreation and vacation by the sea in the summer, and their number is steadily growing (with the exception of the pandemic year 2020).
- Although there are no cultural values of global importance in the destination, the attractions and landmarks, as well as the cultural calendar, retain visitors in the city itself, i.e. individual Bulgarian and foreign tourists.
- The availability of a high-class accommodation base creates opportunities for increasing revenues from overnight stays and attracting solvent tourists all year round for cultural, business, wellness, sports and other types of tourism.
- The seasonal nature of the tourist demand is strongly expressed, and the number of tourists arriving by car and bus is growing.
- The relatively large number of beds per 100 people of local residents and the increasing trend are unfavorable factors for the sustainable development and satisfaction of local residents.
- A contribution to a higher-value experience and attractiveness of the destination Varna can be achieved through the active presence of the destination in social networks and by attracting and servicing tourists through the official tourist website and the TIC.

3.3. Comparative Characteristics of Varna, Burgas, Thessaloniki and Dubrovnik as SSTDs

We compare Varna to cities with similar characteristics and urban growth in order to establish its position in terms of achievements and level of development as an SSTD. Thessaloniki in Greece, Dubrovnik in Croatia and Burgas in Bulgaria are the selected cities according to a number of indicators. The first indicator is the location and size of the destinations similar to those of Varna: medium sized in population and located on the coast. The second indicator concerns the main types of tourism that develop these destinations: marine recreational, cultural, event tourism and other types, mainly during the summer season. The third indicator is economic development and strategy: important port cities with a stable economy and ambitions to be year-round STDs. The fourth indicator is infrastructure and tourist superstructure: access by land (road and rail) and by air (have an airport), developed accommodation and food facilities. The fifth is the presence of a rich cultural and historical heritage and a centuries-old historical past. Summary data for the four destinations are presented in Tables 3.4 and 3.5.

3.3.1 Destination Burgas

Burgas Municipality has 204,804 inhabitants (NSI, 2021) and continues to rank fourth in terms of population in Bulgaria after Sofia, Plovdiv and Varna (according to the 2021 NSI census).

Burgas launched the application of the model SMART CITY (Municipality of Burgas, 2019) following the example of a number of successful European cities, after the municipality joined the Sharing Cities project. In this, the leading cities are Lisbon, London and Milan, while Bordeaux, Burgas and Warsaw are followers that locally apply urban digital solutions and models for the cooperation of leaders (Sharing Cities, 2022). The city-sharing project is proof of a better approach to making smart cities a reality, seeking to develop affordable, integrated urban solutions on a commercial scale with high market potential.

In recent years, the Municipality of Burgas has introduced a number of smart systems for mobility management, waste management, environmental monitoring, disaster risks, the video surveillance of urban spaces and public transport sites. Since November 2019, Burgas has been uniting them in a common integrated platform called Smart Burgas (SmartBurgas, 2023), a project in development that can acquire a key role in improving the quality of life of citizens and their more active involvement in activities related to urban planning and development.

3.3.1.1 *Intelligent control*

SmartBurgas is the digital brain of the city and collects data from various smart devices and systems. The integrated monitoring platform helps the city by bringing transparency, facilitating control and positively influencing the organization of services and significantly improving the quality of

Table 3.4. General comparative information about the destinations Varna, Burgas, Thessaloniki and Dubrovnik (2022–2023).

	Varna	Burgas	Thessaloniki	Dubrovnik
Population	341,737 (NSI, 2022)	204,804 (NSI, 2022)	813,793 (World Population Review, 2022)	42,615 (WeGov, 2017)
GDP per capita (euro)	17,872	15,815	14,273 (Black Sea CBC, 2021)	N/a
Location	Black Sea	Black Sea	Aegean Sea	Adriatic Sea
Port	Second largest in Bulgaria	The largest in Bulgaria	Second largest in Greece	Yes
Number of cruise ships docked in 2022	0 (iNews, 2022)	0 (iNews, 2022)	60 (Greek Travel Pages, 2022) (20 in 2019)	345 (Rogulj, 2022) (492 in 2019)
Airport	Yes	Yes	Yes	Yes
Arrivals (foreigners/ through airports) 2022	• 1,480,000 (Capital, 2023) • From 89 destinations • Predominance of regular flights	• 1,630,000 (Capital, 2023) • From 85 destinations • Predominance of charter flights	• 3,997,858 (SKG Airport Greece, 2023) • From 85 destinations (Thessaloniki Airport, 2023)	• 2,149,181 (Dubrovnik Airport, 2023) • From 127 destinations
Types of tourism	Marine recreational, cultural, event, festival, city breaks	Marine recreational, cultural, event, festival, city breaks	Marine recreational, cultural, city breaks, event, cruise	Marine recreational, cultural, cruise, event, festival
Annual events	Varna Summer International Music Festival, International Folklore Festival, International Theatre Festival, International	Festival of Sand Figures, Jazz in Burgas, Teen Boom Fest, Flora Burgas, MTV Festival "Spirit of Burgas,"	International Film Festival, International Trade Fair, Food Festival, Pride Festival, World Music Expo (MDAT, 2021)	Carnival in Dubrovnik, Summer Festival, Summer Music Festival

(Continued)

Table 3.4.   (*Continued*)

	Varna	Burgas	Thessaloniki	Dubrovnik
	Film Festival "Love is Folly," Golden Rose Bulgarian Feature Film Festival, International Animation Film Festival	National Competition for Pop Song "Burgas and the Sea"		
Participation in projects related to smartness	• Part of the project "Smart and Sustainable Cities" (my SMARTlife) with the CARTIF Foundation as a follower city • Festivallinks Project	• Part of the Sharing Cities project Smart Burgas platform (Burgas Municipality, 2018)	• Intelligent Thessaloniki (2008) • DESTI-SMART project (2021) • Thessaloniki in 100 resilient cities (2015)	• Best Smart city of Croatia, 2016 (Batchelder, 2018) • Respect the city, 2017 (Puljic et al., 2019)
UNESCO World Heritage Site	N/a	The Old Town of Nessebar (near the Municipality of Burgas)	15 monuments in Thessaloniki	Dubrovnik Old Town
Social media presence	• Facebook: 13,000 followers • Instagram: 1535 followers and 535 posts • YouTube (@visitvarna8021): 464 subscriptions and 182 video clips	• Facebook: 30,240 followers and 25,936 shares • Instagram: 2014 followers and 665 posts • YouTube (@gotoburgas): 103 subscriptions and 20 videos	• Facebook: 47,000 followers and 44,000 shares • Instagram: 29,300 followers and 1,742,000 posts • YouTube (@ThessalonikiTravel): 451 views, over 50 videos and 225,041 impressions	• Facebook: 120,000 followers and 119,000 referrals • Instagram: 13,500 followers and 1813 posts • YouTube (@ Dubrovnik Riviera): 79 subscriptions, 10 videos and 418,586 impressions

Table 3.5. Quality of life in the destinations of Burgas, Varna, Thessaloniki and Dubrovnik (2023).

Index (Data from February 2023)	Burgas	Varna	Thessaloniki	Dubrovnik
Index quality of life	149,16 high	138,26 moderate	116,61 low	165,95 very high
Purchasing power	51,65 low	52,63 low	41,05 low	49,20 low
Safety	61,48 high	63,31 high	46,6 moderate	81,87 very high
Health care	53,53 moderate	62,17 high	55,3 moderate	58,43 moderate
Climate	87,37 very high	85,08 very high	88,39 very high	93,24 very high
Cost of living	37,65 very low	41,02 low	58,1 low	52,73 low
Property price/income ratio	8,13 moderate	9,05 moderate	11,75 moderate	17,72 very high
Commuting time	18,27 very low	28,83 low	28,92 low	21,00 very low
Pollution	47,6 moderate	60,15 high	63,85 high	23,45 low

Source: Based on Numbeo (2023).

life of its citizens. Smart Burgas is accessible through a web application from mobile devices with the Android 10+ operating system.

The list of electronic administrative services in the portal of the municipality contains 31 items, and the list for work with the web application of the portal includes the following:

- a specialized web application that provides data from the geographic information system (eGIS) of the city through remote access to specialized digital arrays;
- access to the information page;
- Maps of Burgas, which is an integrated geoportal and a Wireless Internet Places Map;
- a polls and surveys module where residents can give their assessment of urban planning and development;

- information for investors is offered about vacancies in industrial zones, statistics in companies from the business map of Burgas, with location, number of employees and type of activity.

3.3.1.2 *Sustainability*

The Municipality of Burgas works to establish a "green identity" by implementing energy-efficient measures in public transport, using renewable energy sources, and increasing the energy efficiency of buildings and street lighting.

Burgas generates 95,000 tons of waste annually, with 372 containers for separate waste collection and 1,141 for mixed household waste.

The existing infrastructure for the collection and treatment of mixed household waste of Burgas Municipality contains the following elements:

- Regional landfill — cell 2 with a capacity of 400,000 tons with installations for

  - separation of mixed household waste with a pre-separation module for the separation of inert mixtures and crushed plant and biodegradable waste with a capacity of 160,000 tons/year;
  - composting plant waste from the maintenance of green areas (grass clippings, branches, leaves, etc.) with a capacity of 12,000 tons/year;
  - treatment of construction waste from households;
  - crushing large-scale waste;
  - ecopark for temporary storage of hazardous and other specific waste streams;

- Infrastructure for anaerobic treatment of biodegradable waste with a capacity of 30,000 tons/year is being built, intended for the municipalities of Burgas, Nessebar and Pomorie.
- A system for separate collection of biodegradable household waste is being built, including cars.
- Systems for smart management, collection and analysis of a waste-management database are being built and upgraded.

Through a developed software solution, the data on the measured integral noise levels are visualized on the website of the municipality (www.burgas.bg), in the "Environment" section. They are provided by the three municipal stations for continuous noise monitoring. Data from points located in areas exposed to heavy road traffic show noise levels of 9–16 dB(A) above the maximum permissible rate. Persistent noise pollution and increased noise levels are due to the high intensity of car traffic, especially during peak hours of the day and the short distance between buildings and roadways.

Burgas is an important industrial, commercial, transport and tourist center. Some of the industries in the region are unique and defining for the country's economy, such as the production of petroleum, chemical products, plastics, manufacturing of other chemical products, shipbuilding, air ventilation and purification equipment and industry.

The GDP of Burgas District for 2021 was 3,309 million euro (NSI, 2023). In all, 16,083 companies are registered in Burgas. Companies in the private sector are mainly concentrated in industry, construction, transport and commerce. The economy is diverse, which makes Burgas a leader in the southeastern region of Bulgaria. The unemployment rate is 4.2% compared to 9.2% for Bulgaria (Investment Portal of Burgas Municipality, 2023a)

The number of categorized accommodation establishments (over 10 beds) is 830.

In the territory of Burgas District, over 40% of the mass tourism in the country takes place (Investment Portal of Burgas Municipality, 2023b). The Burgas region is an area with an exceptional diversity of natural, anthropogenic and cultural tourist resources.

The oldest buildings in Burgas are the baths built by Suleiman I (1520–1566) and "St. Anastasia" monastery which is located on the island of the same name. As a cultural monument of national importance, it is the only island and the best-preserved monastery on the Bulgarian Black Sea coast.

Near Burgas are the city of Nessebar, a UNESCO World Heritage Site, the ancient and medieval town of Apollonia (today Sozopol) and the ancient Thracian sanctuary Begliktash (near Primorsko).

Burgas municipality is characterized as an urbanized center with medium load.

3.3.1.3 *Hard smartness*

Thanks to a number of smart devices that power the innovative urban platform Smart Burgas with data, residents of and visitors to Burgas can now easily monitor what is happening around them, in real time, in the following key areas:

- *City traffic*: All buses are equipped with cameras and monitor the peace and safety of passengers.
- *Video streaming in real time*: The video surveillance covers the important thoroughfares, pedestrian crossings, schools and busy places in Burgas. It covers all public transport buses and 216 bus stops. Burgas Municipality provides 24-h monitoring in all schools and kindergartens. It monitors the safety of children and reduces the risk of accidents, and by 2023, 41 schools had video surveillance. According to the plan for the integrated development of Burgas (Integrated Development Plan of Burgas Municipality, 2021–2027), more video devices are to be placed in the Seaside Park, around squares, playgrounds and sports grounds, pedestrian crossings and other key places. In this way, the Municipality of Burgas provides a higher level of security to citizens.
- The traffic situation is monitored real time.
- Streets to be repaired can be tracked.
- Objects for sports and health can be monitored.
- Free parking spaces can be identified.
- Rental options for conventional and electric bicycles can be offered.
- Up-to-date data on the quantities of waste collected separately will be available.

In recent years, Burgas has earned the nickname "The Bicycle City of Bulgaria." The city is the first in the country to have a public bicycle rental system (rent-a-bike) and has built over 24 km of bike lanes (0.005 m per 100 m^2, or 4.46 km per 100 km^2), with more coming all the time. The rent-a-bike system Velo Burgas is currently available in 14 locations in the city. The bike stations have different capacities, providing a total of 125 bicycles for rent.

The official tourist portal of Burgas Municipality is https://www.gotoburgas.com. Although it is only available in two languages, the website is extremely interactive, with the ability to share the experiences of tourists and a card (Burgas Card) for discounts at key attraction sites. The portal offers up-to-date information about the weather, transport accessibility, parking opportunities and accompanying services. Interesting routes are presented, oriented to different interests, and have a connection with a special platform for events that are categorized by types and topics.

3.3.1.4 *Soft smartness*

Burgas hosts large-scale musical events, offering tourists an excellent opportunity for a coastal holiday, combined with the dynamics of concerts, exhibitions, theater productions and sports events.

Among the most significant, besides the traditional ones, are the Exhibition of Sand Sculptures, the Stereo Festival for July Morning on the beach, as well as the spectacular party of the Metropolis at Port Burgas, the Spice Music Festival, Jazz in Burgas and the Teen Boom Fest, which presents live influencers and vloggers. At the height of the festival season, the occupancy in accommodation establishments in the Municipality of Burgas is 100%.

The municipality invests in healthcare and innovative social services. A sufficient number of vacancies in kindergartens and nurseries are available in the city unlike other developed cities in the country. Burgas is the first Bulgarian city where English became mandatory in kindergartens, and with the Bulgaria High School for Computer Programming and Innovation, the only one of its kind.

3.3.1.5 *Quality of life*

According to the investment portal of the municipality, Burgas is the most dynamic and fast-growing city in Bulgaria, with a quality of life above the national average (Investment Portal of Burgas Municipality, 2023c). Among the priorities of the city are creating conditions for sports and recreation, as well as extending the tourist season through the development of congress tourism. The municipality pursues a consistent policy of improving the conditions in educational institutions, both in terms of infrastructure and toward the healthier nutrition of children. In addition to the convenient online application system for gardens and schools, through an innovative online platform, you can order and monitor the daily lunch menu of students.

3.3.1.6 *Quality of the tourist experience*

On the integrated platform Smart Burgas, residents of and visitors to the city can find information about the following:

* cultural and historical landmarks;
* current events;

- a list of mobile applications that can be useful in Burgas, such as VR Burgas, Urbo parking (for parking payment in green and blue zones), Burgasbus (for planning routes by public transport) and Evrotrust (for the protection of work with electronic documents);
- whether or not the weather is suitable for the beach in the summer or a walk in the winter;
- data on the levels of water bodies;
- clean air-quality data;
- a set of interactive maps visualizing the numerous objects of interest and allowing for a choice and combinations of themes that can be seen on one screen.

3.3.1.7 *Attractiveness of the tourist destination*

Burgas Municipality is distinguished by a unique combination of natural and anthropogenic resources and transport in a geographical location for the development of tourism: access to the sea, impressive biodiversity and important bird areas, a beautiful urban environment, cultural and historical landmarks, access from the highway and airport, port and railway infrastructure. There is a rich cultural program and seasonal attractions.

The available beds and commercial and restaurant facilities generally satisfy the needs of tourists. The number of accommodation establishments has increased significantly over the years, with 406 (over 10 beds) in 2020, offering a capacity of 6,648 beds. Despite these conditions and despite the general national impetus for the development of tourism, the municipality is not recognized as a tourist destination. The average length of stay is just over 2 days (for foreigners: 2.7 days). The effective use of bed capacity was 37.71% in 2018 and as high as 43.58% in 2014. The average efficiency of the hotel bed base in the country is 39.9%. Importance is given by the higher occupancy of the hotels during the summer season and partial occupancy during the other months of the year. Only five of the accommodation places offered are 4 or 5 star. With the significantly increasing number of accommodation places, tourist tax revenues increased more slowly in the period 2015–2018. The reason for this is the concentration of attendance in the lower-category accommodation places, where the tourist tax is lower.

The Tourism Development Strategy of the municipality for 2018–2023 states that, although they have potential in terms of employment, the means of shelter cannot absorb a significantly increased tourist flow, especially mass visits related to event tourism. Tourists head to the coastal resorts north and south of Burgas, and although Burgas is the gateway to the resorts along the Black Sea coast from a transport point of view (considering the flow that comes from the Trakia highway, the airport and Burgas railway station), Burgas remains a distributor of tourists without benefiting sufficiently from the passing stream. Thus, the available potential remains untapped. These problems are clear and action is being taken. A Tourism Development Strategy has been developed, with clear objectives. Municipal Enterprise "Tourism" (2013) was established to serve municipal tourist sites, complexes and attractions. It manages several sites:

- The Island of St. Anastasia tourist complex;
- The Anastasia tourist ship;
- The Aqua Kalide tourist complex;
- The Clock tourist information center.

3.3.2 Tourist destination Thessaloniki

Thessaloniki had a population of 815,000 in 2023, an increase of 0.12% from 2022. The city of Thessaloniki is a major tourist destination for the north of Greece, with numerous effects on its economy, social cohesion, urban planning and image. Today, Thessaloniki is a vibrant, intercultural city that attracts young people, tourists and international businesses. Thessaloniki is a medium-sized coastal city, located in northern Greece, on the Aegean Sea in the Gulf of Thermaikos, the second largest city in Greece and the second largest export and transit port in the country.

Thessaloniki has been nominated for the Cultural Capital of Europe and the European Youth Capital in different years. The UNESCO World Heritage List includes 15 early Byzantine and Byzantine monuments in Thessaloniki. The Archaeological Museum, the Museum of Byzantine Culture, the White Tower Monument and the fortifications around the city are key landmarks attracting tourists to the city.

3.3.2.1 *Smart governance*

The Municipality of Thessaloniki participates in a Smart Cities Consortium together with the Heraklion Municipality, the Free Software/Open Source Software Society (EEL/LAK), the Technological, Economic and Strategic

Analysis Research Group of the Society of Information (INFOSTRAG) of NTUA, the Association of Mobile Application Companies of Greece (SEKEE) and the Municipality of Athens. As a result, the following smart management initiatives and services are available:

- Open Budget Capture of the implementation of the budget, as it is at the time of access;
- Open Data Online Platform with access to data of the Municipality of Thessaloniki;
- Electronic services (e-services) by the municipal administration;
- "I improve my city," a platform for managing citizens' daily problems, providing functions for submission, management and analysis of citizens' requests;
- Apps4Thessaloniki (Applications for Thessaloniki), a crowdsourcing platform for collecting ideas from citizens, creating online and mobile applications;
- Hackathess, a marathon for application development (hackathon) in order to improve the functioning of the municipality and the city by using the capabilities of ICT;
- Geospatial Information Infrastructure (GIS) at Thessaloniki Community;
- STORM project, providing best practices and applications in the field of cloud computing, in particular when used by public organizations (City of Thessaloniki, 2023).

3.3.2.2 *Sustainability*

Sustainable mobility and accessibility policies are a high priority in the city, but they lack an integrated development plan and mobility management. The mobility system in Thessaloniki faces a number of challenges, including limited opportunities for public transport, excessive dependence on the use of private cars and aging infrastructure. This leads to major traffic jams, with more than 1,600,000 journeys taking place in Thessaloniki every day. Over 52% of the vehicles are private, public transport accounts for 30%, with 8% for other modes of transport. Mobility problems are a major stress factor affecting the daily lives of the inhabitants of Thessaloniki. One of the city's top priorities is the fight against air pollution. The combination of a poor road network, insufficient parking spaces, a small share of public transport, traffic jams and accidents and delays that occur every day make Thessaloniki one of the most polluted cities in not only Greece but also the European Union (Chrysostomou, 2015).

The urban waste management program (which has become imperative) and the implementation of new programs in all forms of recycling (paper, packaging, and glass) in Thessaloniki have been reorganized.

The Municipality of Thessaloniki applies monitoring indicators covering 10 thematic categories, such as education, energy, environment, health and transport, in order to adapt the global Sustainable Development Goals to local conditions, taking into account the international standard the ISO 37120 (sustainable cities and communities) and ELOT 1457. The ELOT 1457 standard has been developed for the needs of Greek cities to support the implementation of sustainability strategies by municipalities, local communities and local partners by calculating 15 sustainability-reporting indicators in the four pillars of sustainability: economy, environment, society and governance.

Thessaloniki accounts for 64.1% of the GDP of the Central Macedonia region (Black Sea CBC, 2021).

3.3.2.3 *Hard smartness*

Thessaloniki participates in the European program Digital Cities Challenge (Intelligent Cities Challenge, 2019). The European Commission's initiative aims to help achieve sustainable economic growth in participating cities by integrating advanced digital technologies. Thessaloniki's ambition is to become a living laboratory for the development of innovative digital services and products. The municipality hopes to support the digitalization of companies focused on activities that are vital to the local economy (e.g., tourism, ICT, wholesale and retail, transport and logistics). As a result of the activities of the Smart Cities Consortium in Greece, a physical/virtual center (Hub) is being created to disseminate innovation, open data and technology and provide integrated solutions in the field of smart environment and smart mobility using ICT.

The "Intelligent Urban Mobility Management and Traffic Control System" (Anastasiadou and Vougias, 2019) to improve the quality of the urban environment in central Thessaloniki is divided into two separate but complementary and parallel units: the Urban Mobility Centre and the Intelligent Traffic Management Centre. The first provides information (accessible via the Internet, telephone and SMS) to travelers, ranging from the provision of tourist information to the planning and execution of the trip. The second focuses on real-time adaptive traffic lights, and its operations include incident management using real-time information, dynamic traffic assessment for future time periods, estimation and validation of predicted travel time and traffic light management in order to reduce the environmental impact of traffic.

Currently, three public transport systems are planned: the metro system is under construction after several delays. In addition, urban maritime transport options are being explored, and a new (western) suburban rail line is planned.

The Earth Observation Toolkit for Sustainable Cities and Human Settlements (2023) in Thessaloniki collects data and measures indicators with a footprint on the geographic information system (GIS) environment. The Geospatial Key Performance Indicators (GeoKPI) and the sustainability strategy of the municipality "Thessaloniki 2030" are in line with the 17 Sustainable Development Goals. The Municipality of Thessaloniki cooperates with the local ecosystem, stakeholders and civil society to plan and implement strategic goals for the city's sustainability and resilience.

Thessaloniki uses the Copernicus space program to upgrade the city's existing spatial data infrastructure (SDI) to provide end users across the Central Macedonia area with advanced satellite data searching, viewing and downloading services. This is done almost in real time through standardized web services, with the new satellite dataset from Sentinel 5P, which gives information on air pollution, integrated into Serial Digital Interface. The scope of this specific mission is to map the global atmosphere every day at a resolution of up to 7 km × 3.5 km. At this resolution, air pollution over the city can be detected.

SDI includes available data on atmospheric concentrations of gaseous pollutants such as nitrogen dioxide (NO_2), ozone (O_3), carbon monoxide (CO), sulfur dioxide (SO_2), formaldehyde (CH_2O) and methane (CH_4) particles (indirect via aerosol index), as well as various information on clouds. The observatory also serves as a repository for other environmental data related to the city's green spaces.

There is a huge development of wired and wireless networks to provide broadband to all citizens. In addition, free Internet access for individual consumers and businesses is envisaged. Sensors have now been installed for real-time information processing and warning, resulting in a smarter environment.

Cycling infrastructure is expanding, but nevertheless, only parts of the city center have a well-established infrastructure of 12 km of cycle lanes and seven bicycle-sharing stations, mainly along the coast.

New business opportunities are emerging. The Shared Bicycle System (Bikethess) and Bicycle Courier Services (PiediVerdi) are examples of innovative businesses creating new or expanding existing mobility practices.

The city lacks an integrated smart mobility tool that could increase access to the various tourist sites through a more sustainable mobility system.

Cross-border accessibility between Greece and Bulgaria is improved through the EasyTrip passenger information system. EasyTrip was developed in cooperation with municipalities from Greece (Thessaloniki, Thermi, Kavala and Serres) and Bulgaria (Bansko and Krumovgrad) and the EKETA and TRAINOSE institutes. The services are provided to travelers through an online platform and also through smart device applications. EasyTrip offers both trip planning and information services (for places of interest, routes, offers, road traffic, etc.), taking into account the user's location and personal preferences.

3.3.2.4 *Soft smartness*

Intelligent Thessaloniki concentrates on the most significant areas of innovation and entrepreneurship in Thessaloniki. The port, business and shopping center, the Aristotle University campus in Thessaloniki, the technology district of eastern Thessaloniki and the airport have been enhanced thanks to a wide range of digital applications and electronic services.

In 2019, the Municipality of Thessaloniki won the "Greek Green Awards" (Geospatial Enabling Technologies) in the category of "Digital Strategy for a Smart City" — the best practices used by local authorities in terms of the environment, sustainability, smart cities and the circular economy.

Thessaloniki was selected as part of the second group of cities to join the 100 Resilient Cities (100RC) network in 2014. The Municipality of Thessaloniki believes that this is a unique opportunity to implement a robust participatory approach to a long-term strategy that addresses current and future challenges. The 100RC methodology provides an innovative model to help local authorities develop a holistic urban strategy in collaboration with neighboring municipalities, local academic institutions, the non-profit sector, private stakeholders and city citizens. The strategy also connects Thessaloniki with other cities and organizations around the world through the 100RC network.

Thessaloniki became a member of IBM's "Smarter Cities Challenge" program (IBM Smarter Cities Challenge, 2015). With IBM's support, the city aims to become smarter by integrating and leveraging a variety of environmental, mobility and management data sources. The Open Data Platform aims to improve communications between the city and its citizens to allow the latter to participate more widely in local decision-making.

The GeoCulture app aims to discover the cultural content of a place. The name GeoCulture indicates the relationship between the cultivation of the land and the cultivation of the mind. The exchange of ideas, values and norms in the context of numerous cultural, geographical and political debates and conflicts is at the heart of the concept (GeoLand, 2023). Within the program, the city of Thessaloniki is hosting the 8th Biennial of Contemporary Art, which takes place from December 2022 to May 2023.

The website of Thessaloniki, part of the VisitGreece.gr website for the whole of Greece as a tourist destination, is extremely interactive, fitting the local into the national identity. The site includes sectors of interest (culture, leisure, religion, landmarks, gastronomy and wine, unique experiences, etc.), an interactive map, a guide, up-to-date information on transport and the opportunity to share opinions and information in social networks. At the same time, there is an official tourist portal of Thessaloniki (www.thessaloniki.travel), which is extremely interactive, with a wealth of up-to-date information in nine languages, opportunities for feedback and sharing experiences, thematic routes and suggestions for tourists of interest.

3.3.2.5 *Quality of life* (*residents*)

The Municipality of Thessaloniki participates in various projects and provides services aimed at "smart citizens":

- *GreenCities*: This is an integrated interactive online game to teach toddlers the methods of rainwater management in Thessaloniki.
- *Composition*: This is an online platform for promoting the action, communication and networking of groups of active citizens.

In the Municipality of Thessaloniki, in order to provide the best services to citizens, a unified system (City of Thessaloniki, 2023) is applied for the operation of the cleanliness services and planned design based on citizens' requests and current conditions in the city.

The Thesswiki project aims to digitize the history and culture of Thessaloniki by the citizens themselves through Wikipedia.

Project "Digitization" is the documentation of cultural documents, with an emphasis on the cultural heritage of the country and its promotion through the display of cultural documents.

3.3.2.6 *Quality of experience (tourists)*

Some of the applications aimed at improving the quality of the tourist experience in Thessaloniki are as follows:

- *The HotThess service*: This uses the "power of the crowd" to offer innovative services to visitors and residents of Thessaloniki. The information is presented in real time, and you can always find out which place or area is worth visiting. HotThess is a service that depends on the data collected by the activity of the social network in the region of Thessaloniki. These data are then used to show the most popular places in the city that locals like.
- *Lark*: This is a platform for the interaction and study of tourism, created with the aim of strengthening alternative tourism.
- *ThessOpenTrip*: This is a smart mobile application aimed at tourists visiting Thessaloniki in order to facilitate their stay. The application uses crowdsourcing techniques to generate data from end users. It allows multiple people to create, evaluate and interact with points of interest (POIs). The app was created to showcase Thessaloniki's attractions to visitors to the city and beyond.
- *CityXplorer*: This is an application that offers the visitor to Thessaloniki the opportunity to get acquainted with attractions, interesting historical places and the way of life of the city.
- *GoTooThessaloniki*: This was created not only for visitors to Thessaloniki but also for citizens who wish to trace the centuries-old history of this wonderful city.
- *Manystories*: This shows that the "stories" of the city are created by the users of the app, who have the decisive say on this.

- *CreativeTourism*: With this, the visitor is given the opportunity to discover the modern face of Thessaloniki and visit the workshops and studios of the artists and get acquainted with them and their work.
- *Tales & Trails*: This is an application that organizes and presents digitally the thematic tours offered by the Thessaloniki Walking Tours team.
- *QuestForThessaloniki*: This is an app that helps tourists organize their visits to interesting places and offers a gaming approach to achieve this.

Building on recent advances in 3D scanning and virtual reality technologies, the e-HOE project, implemented by the Archaeological Museum of Thessaloniki, aims to develop innovative tools for the diagnosis, conservation, documentation and presentation of monuments and sites of the Galerius Palace complex, the most emblematic complex of ancient buildings in Thessaloniki, which includes monuments such as the Rotunda, the Arch of Gallius (Chamber) and the Hippodrome, as well as a virtual connection of the Archaeological Museum of Thessaloniki with the palace complex. The e-HOE tools provide the basis for offering a new multidimensional cultural experience, highlighting two different dimensions: the historical/experiential and the scientific/archaeological through cultural tours in the city.

3.3.2.7 *Attractiveness of the destination*

In 2019, the number of foreign tourists to Thessaloniki exceeded 2 million, which in 2020 dropped to 800,000 and then reached just over 1 million in 2021.

Visitors to Thessaloniki appreciate their experiences in the city in various aspects. The experience with public transport is rated very low (satisfaction ratio 6.3/10), while the average satisfaction rate for the city is 8.1/10.

3.3.3 Tourist destination Dubrovnik

Dubrovnik is a city in Croatia with a population of 42,615. Since 1979, the Old Town has been a UNESCO World Heritage Site. It is one of the most popular tourist destinations in the Mediterranean.

3.3.3.1 *Smart governance*

The smart city Dubrovnik strategy is a basic document of the municipality containing the current analysis of the state of ICT services available in Dubrovnik and plans for the development of projects and technologies for smart cities, published on the city's official website for comments and suggestions. The municipality has implemented five projects in just 2 years (2016–2017), and others are under development.

Some of the projects implemented within the smart community are as follows (WeGov, 2017):

- *Respect the city (Pulijic et al., 2019)*: This is a project that started in 2017 with the aim of managing Dubrovnik responsibly and sustainably.
- *Startup Weekend Dubrovnik*
- *City of Hackathon Dubrovnik*: This involves annual technology events with attention to local engineers, programmers, developers, hackers and manufacturers.
- *Smart City Lab*: This is a common space open to the public, with free Wi-Fi, PC and Macs (integrated company management software), electronic equipment, IoT devices and sensors and 3D printers.
- *Dubrovnik Eye*: This is a web service (Dubrovačko Oko, 2023) with mobile apps for iOS and Android to report utility and similar issues in the city. The citizens' messages are confirmed by the administrator of the city government and allocated to the correct department. Citizens are notified of any change in the status of the problems they note. All data are public and available online.

3.3.3.2 *Sustainability*

In order to prevent air pollution, vehicles are not allowed within the city walls, which is a great attraction for tourists. However, there are not enough parking lots, and the huge flow of buses and cars during rush hour leads to traffic jams. A Park & Ride system has been introduced to alleviate the problem.

- *E-bike*: 15 electric bikes are available at an average price of €30 per day. Bicycles are provided by local shops.
- *Dubrovnik*: This is part of the Plastic Smart Cities initiative and has developed an action plan (Plastic Smart Cities, 2021). In order to

achieve the objectives of the Urban Action Plan, the city of Dubrovnik is working toward fully adopting circular economy models. The Urban Action Plan proposes measures and activities to reduce the use of unnecessary single-use plastics, promotes alternatives to plastic products, designs activities and ensures the implementation of improvements in plastic waste management processes.

The Urban Action Plan consists of four thematic blocks:

1. *Reducing the total amount of plastic waste produced and landfilled, creating a system that supports the circular economy and reducing pollution*: This is achieved by prescribing the objectives and activities to be carried out at the urban level to reduce the volume of waste to minimum legally defined levels.
2. *Creating examples of good practice to reduce waste generation and reuse*: This involves defining the objectives and activities of the city of Dubrovnik to become completely free of plastic, thus being an example of good practice not only in the Republic of Croatia but also in the wider region.
3. *Education and communication with the public*: This involves an activity that fully supports the implementation of the previous two thematic chapters in order to raise citizens' environmental awareness and provide long-term support for the Action Plan to reduce plastic pollution in Dubrovnik 2021–2026.
4. *Dubrovnik*: This monitors and limits the excessive number of visitors (Batchelder, 2018) to various attractions in order to avoid the destruction of cultural heritage.

3.3.3.3 *Hard smartness*

- *Smart parking*: This includes infrared sensor (IR) parking sensors + web + mobile application (iOS and Android) — edge computing implemented at the physical location of the parking lot; the system functions even when there is no Internet connection.
- *People counter*: This is a website with real-time data on the total number of people in the Old Town, six counting cameras in the historic city + 120 cameras at intersections.
- *Dubrovnik*: The official portal of Dubrovnik (visitdubrovnik.hr) presents information in six languages in thematic categories: culture,

food and wine, sports and adventure activities, attractions, events, nature and nearby destinations, accommodation and a link to an app with data on travelers in Croatia.

3.3.3.4 *Soft smartness*

- *Smart city*: This includes mobile applications for the geolocation of users and heat map density of people by combining GPS location data with the total number of people in the city, along with crowd management. They obtain accurate GPS location data of people using smart-city applications, such as Smart Parking, Dubrovnik Eye and Dubrovnik Card.
- *Free Dubrovnik*: This provides access to free wireless Internet in public in the Old City of Dubrovnik. The wireless network has been expanded to include areas, such as St. Dominic Street, the Rupe Museum and the Porporella area. The network is available to all citizens and visitors of the Old Town.

3.3.3.5 *Quality of life*

- Expansion of pedestrian corridors in the city will increase the safety and security of locals and tourists.
- Monitoring and limiting the excessive number of visitors to various attractions will help avoid the destruction of cultural heritage.
- *Intelligent sprayers*: This involves irrigation systems for public parks, with a network of sensors for moisture control connected to a local server for the physical location of the park and humidity mapping.
- *Solar bench*: This includes benches with a modern design, with a canopy to protect against rain, which glow at night thanks to photovoltaic cells that charge with energy, and with a display presenting environmental data: air quality, temperature, humidity, atmospheric pressure and UV radiation and the ability to charge mobile devices.

3.3.3.6 *Quality of the tourist experience*

Dubrovnik is one of the six most visited ports in Europe. In 2017 alone, 639 ships docked in Dubrovnik, with more than 830,000 passengers (Stieghorst, 2017). The number of tourists visiting Dubrovnik per day is 14,000 (WeGov, 2017).

Elite tourism: Dubrovnik is advertised as a holiday destination for the rich and famous, such as Tom Cruise, Roger Moore, Tina Turner and Fatboy Slim. The city actively turns to the rich and famous to cooperate in promoting it as an elite destination. Information about celebrities is published in the *Dubrovnik Times*, a free newspaper in English aimed at tourists. International film festivals and the shooting of television series such as *Game of Thrones* in the city are supported by the municipal leadership and aim to attract more celebrities.

The restriction of bachelorette and stag parties is an additional indication of Dubrovnik's attempts to attract more highly solvent tourists.

In the Old Town, the hotel capacity is about 16,500 beds in hotels, hostels, campsites and homestays (B&B, guest houses, villas, apartments and rooms), while in the Dubrovnik–Neretva region, of which the city of Dubrovnik is the head, there are 62,000 registered beds (Carić, 2011). This region generates potential visitors to Dubrovnik in the form of excursions, but statistics on 1-day visits are lacking. The total number of beds is therefore 78,500. Taking into account the population of the city of Dubrovnik, it turns out that there are approximately two beds per person. This high ratio, as well as the large number of tourists from cruise ships coming without an overnight stay, determines the problem associated with overtourism, especially during the summer months, and the need for crowd control.

Dubrovnik's regional government has actively stimulated hotels and private investors to build new accommodations. An obstacle for investors is the strict restrictions and the non-admission of high-rise construction in order to preserve nature and comply with the principles of sustainability. The management of the municipality of Dubrovnik does not want the city to experience the negative effects of mass tourism, as observed in other destinations. At the same time, the municipality encourages the local population to provide rooms for tourists as financial support for the family budget.

3.3.3.7 *Attractiveness of the destination*

Tourism is one of the fastest growing sectors, with a record increase of 17% in tourist arrivals in 2017 (compared to the overall increase in tourist arrivals in the country of 13%). This creates challenges for visitor management, with most visits occurring during the summer months, from June to September (an average of 4 million overnights during the summer season). In 2019, the number of tourists reached 1.5 million, falling sharply in 2020 to 518,000. But, these statistics do not take into account the number of daily visitors as tourists and passers-by, as well as tourists from cruise ships.

The most visited sights are the Old Town, the city walls and numerous fortresses that are concentrated in a small territory. This leads to crowding, traffic jams and growing discontent. On the contrary, film-induced tourism gives impetus to the development and preservation of Dubrovnik's culture, but at the same time threatens its reputation as a desirable and valuable destination to visit or live in.

Dubrovnik's motto is "a city for all seasons" (*Croatia Week*, 2013). Achieving this goal will mean permanent jobs and a more even tourist flow and cash inflow into the city. Therefore, Dubrovnik strives to provide more winter events and to work closely with low-cost airlines from all over Europe.

In this regard, the Tourist Council of Dubrovnik has created a "Winter Card" offering a program of cultural events, free entrances to landmarks as well as discount vouchers for other attractions and restaurants. Nearly 100 local businesses support this initiative. The offer is valid for tourists who stay for two or more nights or during the winter period.

The Dubrovnik Card application is a mobile application (iOS and Android) for tourist information and not only presents the city's attractions, including those that are nearby, through iBeacons, but also allows discounts in restaurants, retail outlets and others.

The tourist destinations Varna, Burgas, Thessaloniki and Dubrovnik were compared on the SSTD assessment indicators (see Table C.1 of Appendix C). As a result of the comparative analysis, the position of Dubrovnik at a high level as an STD, with a clear social orientation, a smart ecosystem and smart governance, stands out. Thessaloniki is at an average level with achievements in terms of hard smartness and smart management. The two cities still have problems with sustainable development, despite the good practices introduced, especially in terms of monitoring and controlling the number of visitors to the destination and attractions, intensive tourist flow during the summer season and traffic. These are the factors that negatively affect the attractiveness of the destination and the valuable experience for tourists, and in Thessaloniki, the quality of life. At the same time, the measures and tools implemented by local authorities are aimed at increasing the value of smart ecosystem indicators to the level of Valencia and Nice, which are high for an SSTD.

The results of the comparative analysis are presented in Fig. 3.1.

Varna and Burgas are at a basic level as smart tourist destinations. Burgas has better results that meet the average level in terms of the

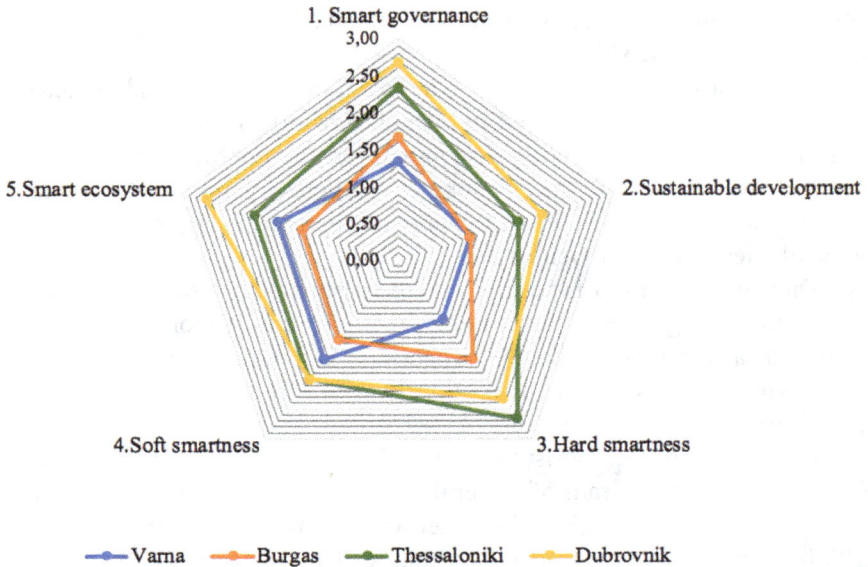

Fig. 3.1. Comparative assessment of Varna, Burgas, Thessaloniki and Dubrovnik as SSTDs.

development of hard smartness and smart control. They are most distinctive for indicators, such as open access to large databases and the ability of citizens to monitor the implementation of plans and sustainable development. In both destinations, the potential for increasing the attractiveness of the destination and the quality of the tourist experience has not been realized.

3.3.4 Conclusions from the assessment and comparative characteristics of tourist destination Varna as an SSTD

3.3.4.1 *Key achievements and potential opportunities*

Some key achievements and opportunities include the following:

- expansion into the creative and ICT sectors;
- attractiveness of the destination determined by a high-category base, cultural institutions and values, traditional and innovative events and geographical location;

- research potential related to the blue economy (maritime industry, maritime transport, port and offshore extraction, fisheries and aquaculture, maritime tourism), medicine and technology.

3.3.4.2 *Features and areas for improvement*

Some key features and areas for improvement are as follows:

- public–private partnership and active stakeholder networks;
- digitalization, which includes infrastructure, business, attractions, service to tourists and citizens, and innovative tourist products;
- sustainable development, which includes support for disadvantaged people, regulatory and incentive measures, engaged business, citizens and tourists, low-carbon and circular economy and innovation for the Sustainable Development Goals.

3.4. Strategy for the Development of Varna as an SSTD 2030

The strategy for the development of Varna as an SSTD 2030 is determined by the key potential opportunities and achievements, as well as the necessary actions to improve directions and characteristics that slow the growth rate and do not allow the creation of added value for the destination as a whole: quality of life, quality experience and competitive business and attractiveness of the destination. In this aspect, a strategy is proposed that covers a general, long-term guideline, supported by two specific, medium-term development directions (see Fig. 3.2). They are committed to taking action and achieving measurable results in the short term. The main purpose is that by 2030 the criteria used to assess the tourist destination as smart and sustainable will meet those for added value and not less than 50% of the exclusivity criteria.

In the long term, the strategy outlines the development of a smart ecosystem for tourism, the creative sector, ITC, education and science as key and fundamental. This ecosystem is an environment that provides opportunities for conducting competitive business and achieving quality of life through knowledge, technology and innovation with no negative impacts on the environment.

The construction and functioning of the smart ecosystem is determined by the development of tourism, which is integrated into the local

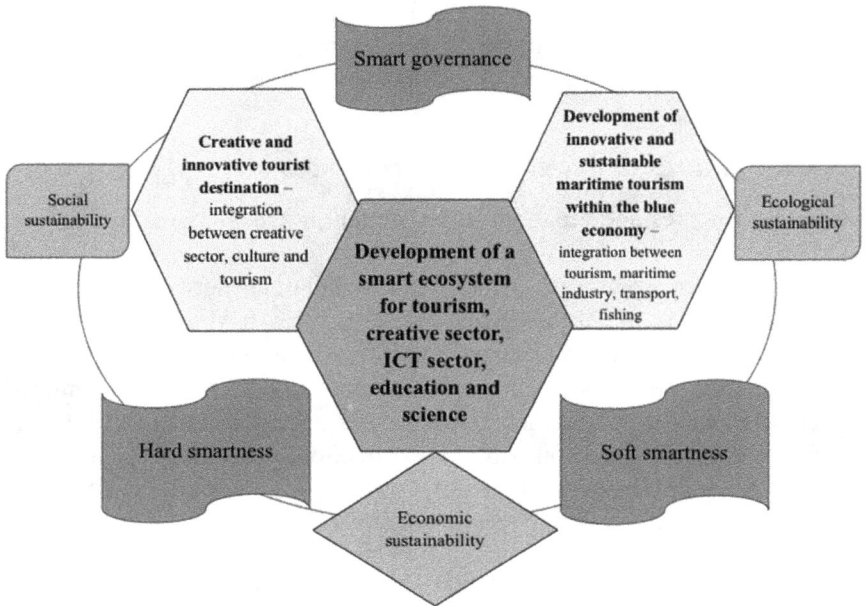

Fig. 3.2. Strategy for the development of Varna as an SSTD 2030.

economy, creates added value for the destination and preserves and enriches its potential through its resources, values, traditions and social capital for future generations. This requires that the research and educational potential in the blue economy, medicine and digital technologies be linked to tourism through cross-sectoral initiatives and policies.

In the medium term, the strategy covers the development of the following:

- *Innovative and sustainable maritime tourism within the blue economy*: This involves the integration of tourism, the maritime industry, fisheries and transport.
- *A creative tourist destination*: This involves the integration of tourism and the cultural and creative industries, regarding cultural, cognitive, creative and event tourism.

Innovative maritime tourism is carried out through the sustainable use of coastal and marine resources, including mineral and geothermal waters, based on research achievements and projects, ecological maritime transport and digital technologies developed and applied in the blue economy and medicine. The areas of interaction include underwater archaeology; travel and entertainment by boat and on the high seas; balneological, spa and healing procedures for citizens and tourists; traditional fishing and pesca-tourism and informal education (summer academies, business accelerators and hackathons).

A creative tourism destination is developed through a partnership between digital, new media, contemporary art, tourism and entertainment, cultural institutions and traditions. The ICT sector has the potential to be used for the digitalization of tourism and creation of products based on high technologies and preserved cultural values. The areas of interaction include, for example, art and cultural events in traditional and non-traditional spaces; cultural, cognitive and entertainment routes; creative communities in food, art, design and archaeology; digital media and contemporary art. Also important is the active and attractive presence of the destination in social networks.

In the short term, the strategy covers the achievement of actions and concrete results in terms of smart management, sustainability, and hard and soft smartness.

3.4.1 Smart governance

- *Initiation and creation of public–private partnerships and partnership networks for the implementation of research projects*: This includes the development and application of mobile applications and virtual assistants; tourist products — routes (digital and assisted with QR codes and RFID), events, offshore trips and underwater expeditions; operation of TICs — real and digital; building and maintaining platforms; database and cloud services for the needs of business, citizens, the public sector and tourists, and for sustainable development.
- *Monitoring of achievements and results*: This involves the implementation of strategies and integrated development plans, criteria and indicators for sustainable development; environmental protection, reducing impacts on climate change, investment climate, etc., mainly

through a shared database and publicly available platforms and applications in which to concentrate the information to be updated in real time.

- *Dialog between business, citizens, tourists and management of a tourist destination*: This is achieved through mobile applications and platforms for sharing information, news, data, suggestions, opinions and assessments, and virtual rooms for regular meetings and discussions.
- *Increasing the number and quality of electronic municipal services*: This is achieved by improving the existing platform, including by adding virtual assistants, and its accessibility to achieve greater efficiency.

3.4.2 Sustainability

- *Eco-friendly transport*: This includes upgrading public transport with environmentally friendly vehicles or an alternative source of energy; improvement of a cycle-lane network and introduction of a cycle rental and sharing system; tourist traffic from airport to resorts and development of sea transport, including with a certificate of environmental friendliness; increasing the intermodality of public transport, including to tourist attractions outside the city, and the introduction of measures to reduce noise emissions.
- *Accessibility of attractions and cultural institutions*: This is achieved by providing access (physical, virtual, linguistic, etc.) to museums, galleries, libraries, cultural spaces, beach, sights and attractions, including natural and entertainment parks, for representatives of socially vulnerable groups: people with disabilities, representatives of minority groups, disadvantaged youth, etc.
- *Certificates for environmental management in tourism*: This involves, at the destination level, including carbon-footprint calculations for tourism organizations, in the blue economy (maritime transport, fishing, etc.), implementation of projects with the participation of research and educational institutions, representatives of sectors of the blue economy, the municipality and the non-governmental sector.
- *Circular economy*: This involves connecting tourism with the production of ecological fuel (use of waste) and use of coastal and marine resources; the production of energy, food, etc., taking into account the environmental impact.

3.4.3 Hard smartness

- *Digitalization of attractions and landmarks*: This involves creating smart attractions by using AR, VR and MR and artificial intelligence to visualize past activities and missing elements, transfer to other worlds, fun through play and challenges for the imagination.
- *Digitalization of services to tourists and citizens*: This involves facilitating mobility/mobile applications for shared transport, use of bicycles, development of (eco-friendly) routes, smart elements of the urban transport environment, up-to-date and easily accessible information about destinations and entertainment, including digital routes, and a digital center for utility and traffic management.
- *Digital spaces for business from tourism, blue economy, cultural and creative sectors, research and training*: This involves the development of a common infrastructure (creation of free high-speed Wi-Fi hotspots) and platform, including implementation of plans for the establishment of a High-Tech District, a Mathematical–Technological Corpus and an Industrial Zone Aksakovo.

3.4.4 Soft smartness

- *Development of the creative sector*: This includes the initiation and support of innovative and creative projects, construction of new cultural and creative spaces, including for contemporary art and new creative communities; creative workshops, places for events and exhibitions, development of "Talyana" (the art and knowledge district in the Old Town of Varna, which has the highest concentration of cultural and historical heritage).
- *Creative tourism practices*: This includes partnerships between cultural institutions (museums, galleries and libraries) and creative communities for events and activities, with the active participation of audiences and tourists.
- *"Live Lab"*: This includes a digital open database and platforms for open innovation, interaction through ICT, a physical environment in which they are united, and an opportunity for interaction between science and the creative and ICT sectors, cultural institutions, education and business; a separate center in the Technology Park, Knowledge District of Talyana or in a new cultural space in the area of the Marine Station is provided.

Conclusion

The establishment of a tourism destination as smart and sustainable is essential for its future development, promotion and successful positioning. Adding value to the tourist experience and improving the quality of life of local residents are priorities when building the smart and sustainable tourism destination (SSTD) concept. This condition requires the development and implementation of a socially oriented strategy for an SSTD. This monograph aims to present and analyze a theoretical model for the preparation of a strategy for SSTDs and to test it for the Municipality of Varna so that the model can be replicated in other territorial units. The structure and content of the development correspond to the main objective and research tasks, and follow the logical sequence of the stages of scientific research.

In Chapter 1, the concept of a smart tourism destination (STD) is derived in the context of its evolution from the concept of a smart city. First, based on an analysis of the definitions, the main pillars, results and goals of a smart city are defined. Second, the conditions and factors analogous to a smart city that determine the need for the reorientation of the development of a tourism destination in the direction of a smart version are described. Third, the definitions of an STD in three directions — foundation, processes and results — are analyzed, and the necessary technologies for its development are summarized. On this basis, an author's definition of an STD is derived. Then, an overview of models and methodologies of scientists and specialists in the field of STD research is made. The principles of sustainable development of a tourist destination and the indicators for its assessment and measurement are discussed and summarized. The interrelationship between sustainable development and STD development is also examined. Based on the theoretical justifications, an author's model for an SSTD has been prepared, which is based on the existing approaches to the development of STD by linking its goals with the principles of sustainable development and the social aspect of their achievement. The model includes smart management, which initiates and creates an intelligent tourism ecosystem, in a sustainable environment, to achieve a higher quality of life for locals and the tourist experience, as well as affirms a higher degree of smartness on the pillars of the destination's attractiveness.

Chapter 2 presents the methodology and toolkit for research and evaluation of SSTDs. The five main indicators are outlined, and a detailed description of the object of study — the tourism destination of Varna — is provided. The study presents the parameters of a survey conducted among three main groups of participants in the process of establishing a destination as sustainable and smart: experts, tourists and citizens. The aim of this study is to determine the level of development of Varna as an SSTD based on the opinions of the representatives of these stakeholders. Opinions are united around the conclusion that the main problems are related to achievements in terms of indicators of mobility, accessibility, digitalization and sustainable development.

Chapter 3 presents the examples of successfully developed and established SSTDs: the cities of Valencia and Nice. Good practices are addressed by using the main groups of indicators of an SSTD. The methodology for the evaluation of SSTDs is applied in the study for an in-depth analysis of the tourism destination, Varna and an evaluation of the cities of Burgas, Thessaloniki and Dubrovnik as similar destinations. The conclusions show the degree of development of Varna as an SSTD, which is defined as a baseline with good potential, and the key achievements and opportunities, characteristics and areas for improvement are outlined. On the basis of the analysis, a proposal for a conceptualization of a strategy for the development of Varna as an SSTD is outlined. It covers a general, long-term direction that is supported by two specific, medium-term development directions: being committed to taking action and achieving measurable results in the short term. In the long term, the strategy is aimed at building and functioning a smart ecosystem for tourism, the creative sector, ICT, education and science as key and fundamental. This process is related to the development of innovative and sustainable maritime tourism within the blue economy and a creative tourism destination. This is achieved through specific activities and achievements in smart management, sustainable development, and hard and soft smartness.

All working hypotheses have been proven through theoretical, methodological and empirical research. The confirmed hypotheses contribute to the adoption of the main thesis presented in Chapter 1 that the development of an STD 2030 is possible if it is carried out in close interrelation with sustainable development and is socially oriented.

This work is the first of its kind in examining the Municipality of Varna through the prism of its development as an SSTD. It stands out with the following main contributions: A model for an SSTD 2030 has been developed; a methodology has been created to assess the degree of development of a tourism destination as smart; an author's definition of STDs has been formulated; good practices and opportunities for the development of Varna as an SSTD have been explored, and guidelines and opportunities for its development are summarized.

The authors do not claim to be exhaustive regarding the presented problems. For example, in the empirical study, only the respondents' opinions against the SSTD criteria were evaluated without analyzing the motives leading to these results. Therefore, in order to determine the reasons for the differences in the assessments of citizens and tourists of the Municipality of Varna according to one of the most important criteria — "significance of cultural heritage and creativity" — it is necessary to carry out additional research in future to introduce a more precise formula to calculate a global index for SSTDs and for each of the areas. At a later stage, it is also possible to carry out research and testing of the proposed model in other SSTDs, located outside of Europe and/or at different stages of their development, thus ensuring the universality of the proposed author's model.

The future trends in the study and development of STDs can be outlined as follows:

(1) connection building between strategy orientation and implementation for developing SSTDs and economic, ecological and political resilience of tourism and destination management;

(2) smart development of a tourism destination based on strategic approaches to the application of AI in destination and urban management, as well as principles of the circle economy;

(3) smart development of a tourism destination parallel to upgrading competencies of the workforce in the context of digitalization, green transformation and the formation of intercultural living and the working environment.

Appendix A: Models and Methodologies for STD Research

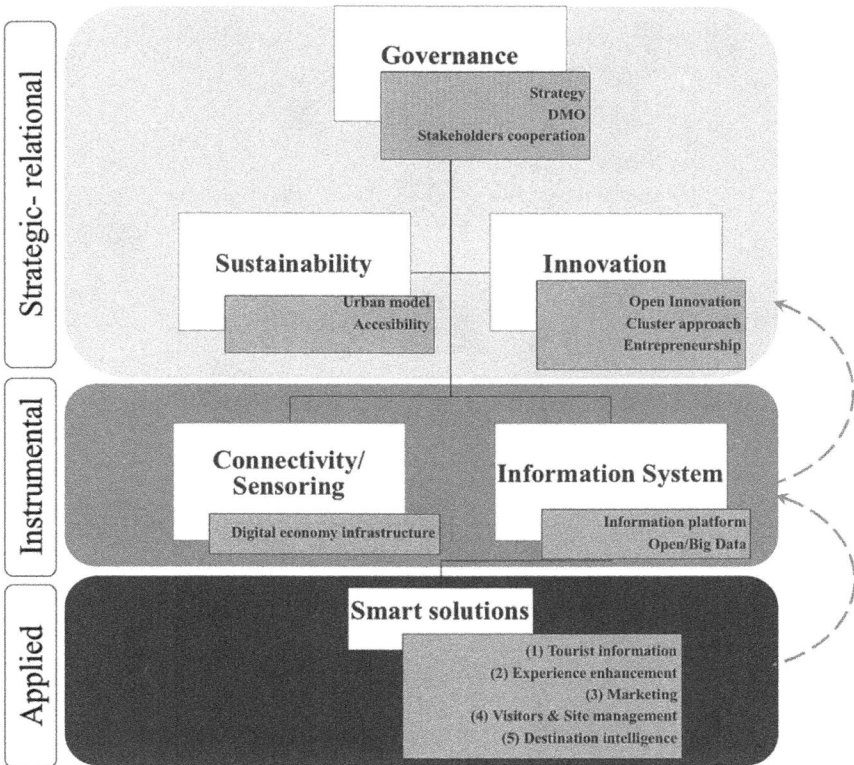

Fig. A.1. System model of a smart tourism destination.

Source: Adapted from Ivars-Baidal *et al.* (2016).

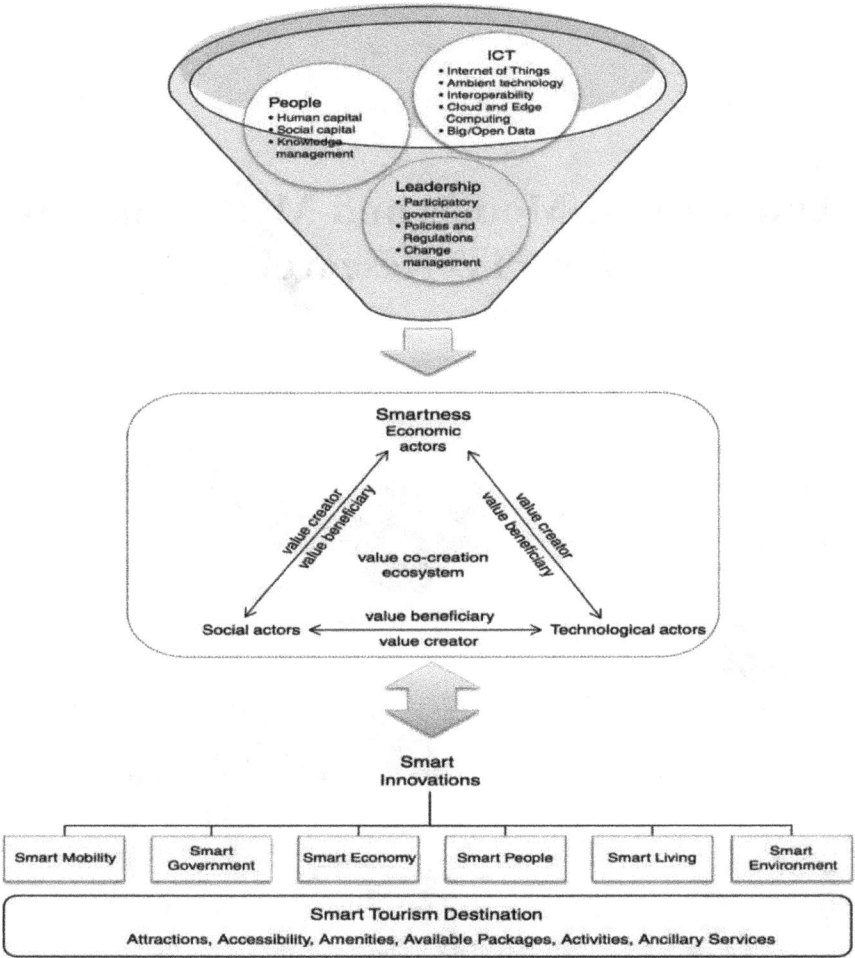

Fig. A.2.   Smart tourism destination framework.

Source: Adapted from Boes *et al.* (2016).

Table A.1.   STD study areas and categories.

Scope	Initiatives and actions studied	Scope	Initiatives and actions studied
(1) Sustainability of the destination and quality of life of its inhabitants	1.1. Water management 1.2. Energy 1.3. Pollution (air, noise, light, water, soil) 1.4. Mobility and transport 1.5. Territorial and urban planning 1.6. Citizens' quality of life situation and improvement plans	(2) Accessibility of the destination	2.1. Physical accessibility of tourist sites 2.2. Digital accessibility of tourist information 2.3. Urban general accessibility
(3) Technological solutions applicable in tourism, connectivity and sensor networks	3.1. Applications 3.2. Websites 3.3. Social media 3.4. Wi-Fi networks 3.5. Sensors 3.6. Augmented reality 3.7. QR codes, RFID[1] 3.8. Interactive platforms	(4) Innovation, management, smart/ information systems applied in tourism	4.1. Local plans related to smartness 4.2. Included public organizations 4.3. Information systems and boards for tourist intelligence 4.4. Management and collaboration platforms 4.5. Open data and transparency

Note: [1]Radiofrequency identification.
Source: Femenia-Serra and Perea-Medina (2016).

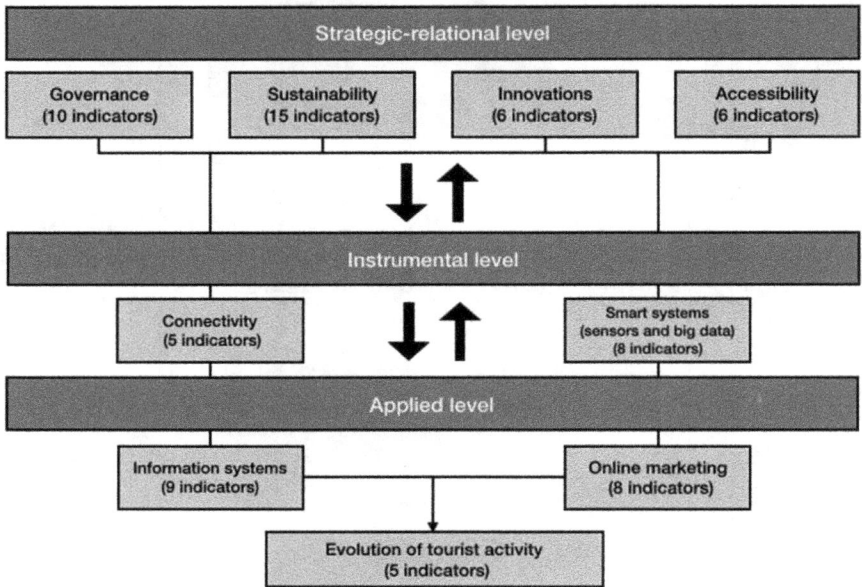

Fig. A.3.   Structure of model of intelligent destination: Field and criteria.

Source: Ivars-Baidal *et al.* (2021), summarized and supplemented by the authors with indicators by criteria.

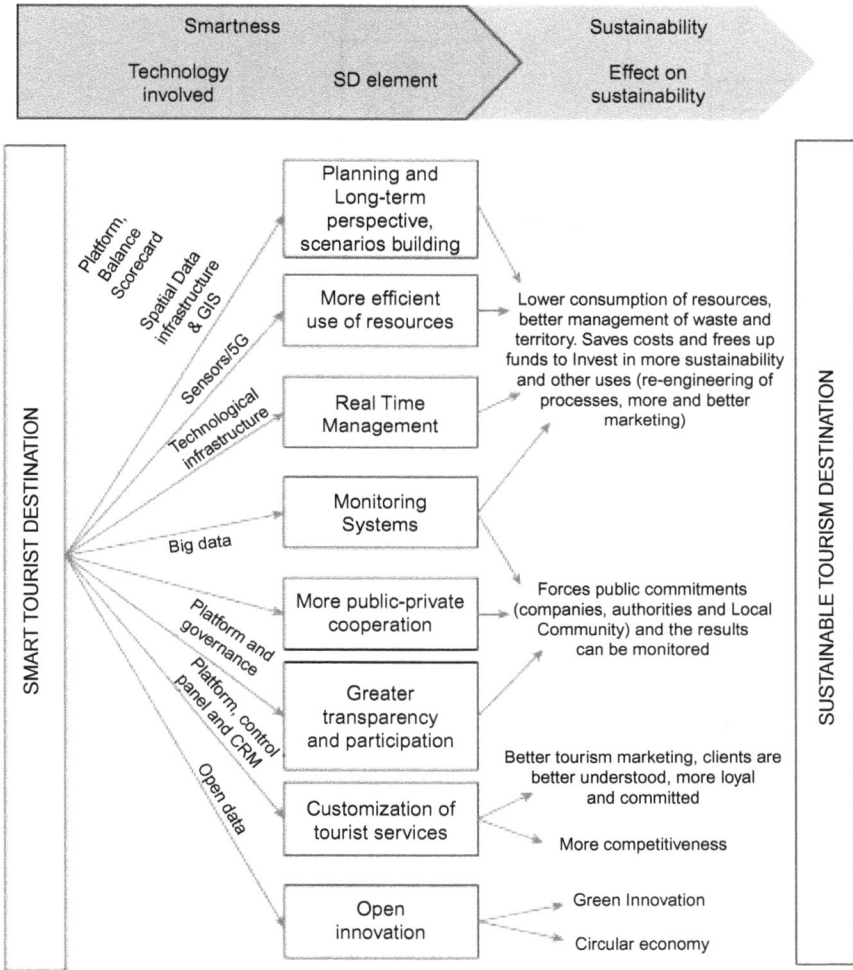

Fig. A.4. Theoretical directions of transforming smartness into sustainability.

Source: Adapted from Perles Ribes and Ivars-Baidal (2018).

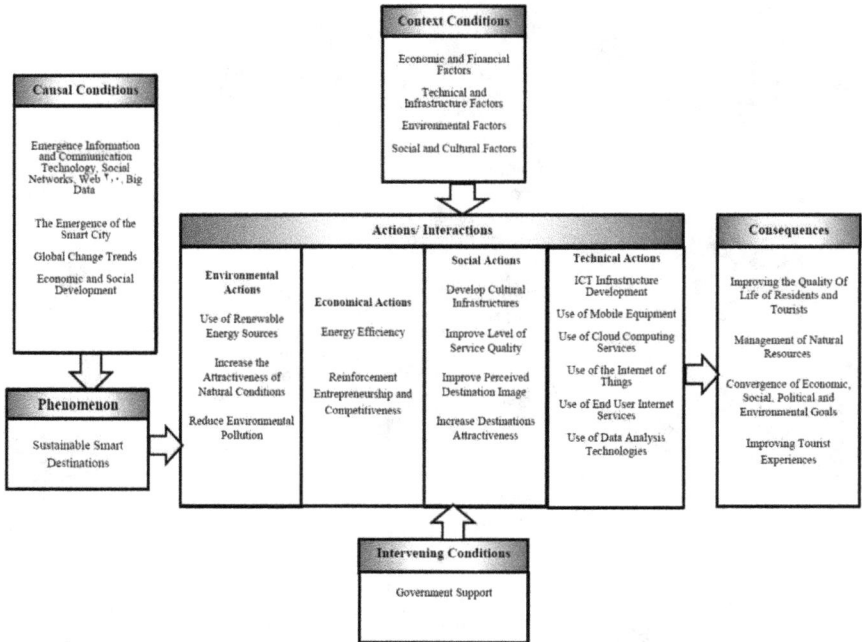

Fig. A.5.   A sustainable smart tourist destination model.

Source: Adapted from Shafiee *et al.* (2019).

Appendix B: Indicators for the Assessment of an SSTD

Table B.1. Baseline indicators for the assessment of SSTDs.

Area	Group	Subgroup	Indicator
Smart governance	Digitalization	eGovernment	Digital, online accessible systems for administrative services: at least five types, one of which is for tourism
		Governance platforms	Publicly available documents related to destination management
	Orientation toward sustainable development	Strategies and plans for sustainable development of tourism and destination	Strategies and plans for sustainable tourism/destination development over at least a 5-year period, publicly discussed and accessible
		Strategies and plans for smart development	Integrated development plans and other plans related to education, ICT, etc., for at least a 5-year period, publicly discussed and accessible
	Orientation toward democracy	Cooperation and stakeholder involvement mechanisms	Institutions: standing committees, councils, departments in municipal structure
			Regulatory regulators: regulations, decisions, acts of local government
		Transparency of local government decisions	Public access to decisions

(Continued)

Table B.1. (*Continued*)

Area	Group	Subgroup	Indicator
Sustainable development	Environmental sustainability	Environmental Impact Management	Presence of double water treatment
			Ratio between biological and mechanical treatment of WWTP
			Availability of a separate waste collection system
			Possibilities of using alternative modes of transport infrastructure, e.g., bicycle, pedestrian
			Facilities and measures to reduce air and noise pollution: protective walls, greenbelts, limiting access of cars, availability of parking spaces, etc., in certain areas
			Share of alternative energy in destination: for production, lighting and heating
			Number of beds and accommodation places per unit area
		Protected areas and resources	Proportion of protected areas and areas with special status out of the total area
			Size of protected areas per capita
			Protected natural resources: number and types
	Social sustainability	Social integration	Accessibility of socially vulnerable groups to accommodation and attractions
		Preserved identity	Applied measures for the preservation of cultural heritage: institutional, regulatory, financial
			Preserved traditions and traditional events
	Economic sustainability	Tourism contribution to economic development	Sustainable and growing share of tourism in GDP, employment, income, investment

	Energy and resource efficiency	Declining coefficients per kW, 1 l of water and others per night, per guest, tourism revenue, GDP
Hard smartness	Digitalization of infrastructure	Internet coverage of the territory Share of public territory with free Wi-Fi access
	Smart technologies in tourism	Interactive destination website Mobile application for tourists QR codes of tourist attractions VR and AR in key attractions Navigation and geolocation
Soft smartness	Creativity	Growing share of jobs/companies in creative industries Events for contemporary and street art and new media Creative tourism products
	Community intelligence	Share of people with university education in ICT, and high-tech firms: similar or above national average Level of digital competencies: above the national average
Smart ecosystem	Quality of life	Developed public transport
	Mobility	Accessibility of socially vulnerable groups to institutions, public places, public transport
	Accessibility	Developed street network, park and public spaces, 100% covered area with water supply and sewerage network
	Quality of infrastructure	

(Continued)

Table B.1.   (*Continued*)

Area	Group	Subgroup	Indicator
	Quality of experience	Satisfaction of tourists	Destination rating: in the top 100 for Europe or the top 10 in the region
		Image of the destination	Share of tourists willing to return again
			Positive destination associations
			Accessibility to the destination by public transport: two types
			Accessibility to attractions and accommodation places: public transport and other means (bicycle, electric scooter)
			Number of tourists and inhabitants per unit area
	Attractiveness of the destination	State of attractions	Growing or sustainable number of tourists year round
			Insured access, opportunity to visit
			Variety of attractions
		Quality of accommodation	Share of properties above 3 star
			Average rating of accommodation in the destination
			Overnights per unit area
			Number of beds per 100 people of the population
	Business competitiveness	Development of the business	A positive rate for development: income, workplaces
			Sustainable or growing life expectancy of the business

Table B.2.   Indicators for assessment of SSTDs.

Area	Group	Subgroup	Indicator
Smart governance	Digitalization	Open databases	Tourism, economy
		Platforms for governance	Publicly available documents and decisions, with the possibility of comments and suggestions
	Orientation toward sustainable development	Strategies and plans for smart development	Digitalization strategies for at least a 5-year period, with specific local government commitments
		Monitoring sustainable development of tourism and the destination	Annual reporting of values according to sustainable development indicators and measures taken to achieve desired values: publicly available
	Orientation toward democracy	Mechanisms for cooperation and participation of stakeholders	Projects to stimulate the participation of stakeholders
		Transparency of decisions	Public access to decisions, decision-making process and implementation
Sustainable development	Environmental sustainability	Management of environmental impacts	Proportion or number of tourist companies connected to WWTPs for biological wastewater treatment (own treatment plants) twice treating wastewater of a pool used for recreation
			Share of electric transport in general transport
			Share of electric cars in total number of cars
			Share of tourist companies offering tourist products with pedestrian/bicycle/electric transport

(Continued)

Table B.2.   (*Continued*)

Area	Group	Subgroup	Indicator
			Share of tourist companies with separate waste collection
			Proportion of tourism companies implementing systems for controlled water use
			Areas used by tourism
		Circular economy and green investment policy	Certificates for environmental management at destination and business level
			Projects and financial mechanisms to stimulate tourism business included in circular economy
			Share of investments in "green" technologies in tourism
	Social sustainability	Social integration	Access of people from socially vulnerable groups to events, entertainment, and education and jobs in tourism
		Preserved identity	Tourist products and services with preserved cultural heritage
	Economic sustainability	Contribution of tourism to economic development	A stable and growing share of revenues (or only revenues) from tourism companies using smart technologies
		Energy and resource efficiency	Investments in energy-saving and resource-saving technologies in tourism
Hard smartness	Digitalization of infrastructure		Using sensors to monitor traffic, weather, visiting public places and attractions
			Over 50% of public territory and places with free Wi-Fi access

Smart technologies in tourism		Mobile applications for tourists, including digital travel guides
		QR codes, VR and AR: wide application in tourist destinations, including in shopping centers, public places, transport
		Navigation and geolocation
Soft smartness	Creativity	Share of study programs and disciplines for culture and creative industries
	Community intelligence	Share of jobs in tourism related to smart technologies and special knowledge
	Innovations related to sustainable development	Tourism companies/social entrepreneurship projects
Smart ecosystem	Quality of life	Intermodal transport (connection between at least two types of transport)
	Mobility	Mobile applications and platforms to access information in real time: transport, traffic, accessibility to attractions, cultural institutions and public places, parking
	Accessibility	Digitalization of educational services
	Quality of infrastructure	Environment- and landscape-friendly infrastructure
	Satisfaction	Satisfaction with tourism development in a destination

(Continued)

Table B.2.   *(Continued)*

Area	Group	Subgroup	Indicator
	Quality of experience	Satisfaction of tourists	Destination rating: above average for similar destinations
			Proportion of tourists willing to return again: above average for similar or other destinations
		Destination image	Predominant share of positive reviews on social media
			Positive destination associations related to smart, modern, sustainable tourism
			Accessibility to the destination by public transport: three types
			Intermodal transportation for accessibility to attractions, shopping and transportation centers and accommodation places
	Attractiveness of the destination	State of attractions	Smart maintenance of attractions: preservation, presentation
		Quality of accommodation	Places of accommodation that implement smart technologies
			Availability of discount cards in various sites, tourist companies, attractions with prepayment
	Business competitiveness	Business development	Share of investments in innovation and smart technologies to serve tourists before, during and after a visit
			Participation in affiliate networks

Table B.3.   Indicators for assessment of SSTD for exclusivity.

Area	Group	Subgroup	Indicator
Smart governance	Digitalization	Open databases	Tourism, economy, transport, ecology, demography
		Platforms for management and collaboration	Platforms for creating partnerships, joint projects and activities, for dialog and exchange of knowledge and experience
	Orientation toward sustainable development	Monitoring sustainable development of tourism and the destination	Institutionalized, with stakeholder participation
		Strategies and plans for smart development	Strategies for innovation and smart growth for at least a 5-year period, with specific commitments of local government
	Orientation toward democracy	Public–private partnerships engaged in the development of a tourist destination	Partnerships for digitalization, cultural programs, urbanization, destination marketing
Sustainable development	Environmental sustainability	Management of environmental impacts	Share of tourism companies using alternative energy
			Share of tourism businesses implementing climate change impact reduction schemes or carbon-footprint calculations
			Implementation of schemes and measures to reduce carbon emissions in the destination, implemented by a wide range of stakeholders and local residents

(*Continued*)

Table B.3.   *(Continued)*

Area	Group	Subgroup	Indicator
	Social sustainability	Policy to stimulate creative industries and social entrepreneurship	Projects, financial mechanisms, strategies
	Economic sustainability	Contribution of tourism to economic development	Investments in tourism related to smart technologies
		Policy to stimulate SMEs in tourism	Support for adaptation to climate change requirements
			Support for digitalization
Hard smartness	Digitalization of infrastructure		Use of artificial intelligence and cloud services for monitoring and support, including for the purposes of sustainable development
	Smart technologies, oriented and contributing to sustainable development		Information systems and platforms for traffic and crowd control, including redirection of tourist flows, restriction or suspension in case of exceeded norms or danger to the environment
Soft smartness	Creativity		Centers and clusters for creative entrepreneurship, "living laboratories"
	Innovations related to sustainable development		Contribution of tourism to the circular economy: number of companies, companies with partners
			Green innovations in tourism
			Mobile applications to calculate carbon footprint with different means of transport at the destination

Smart ecosystem	Quality of life	Mobility	Intermodal transport: all types of transport
		Accessibility	Digitalization of healthcare and social services
			Mobile applications for finding parking spaces, for crowd notification
			Participation of local people in creating tourism products
	Quality of experience	Satisfaction of tourists	Digital tourist guides and thematic routes
			An interactive, bookable destination platform
	Attractiveness of the destination	Quality of additional services	Cashless and contactless payment in tourist sites and attractions when using various services
	Business competitiveness	Business development	Joint tourism products and activities

Appendix C: Comparison of Cities Based on SSTD Evaluation Indicators

Table C.1. Comparison of Varna, Burgas, Thessaloniki and Dubrovnik based on SSTD evaluation indicators.

Indicators	Varna			Burgas			Thessaloniki			Dubrovnik		
	A	B	C	A	B	C	A	B	C	A	B	C
1. Smart governance	2	2		1	4			4	3	2		6
• Digitalization	✓				✓				✓			✓
• Orientation toward sustainable development	✓			✓				✓			✓	
• Orientation toward democracy		✓			✓			✓				✓
2. Sustainability	3			3			1	4			6	
• Environmental sustainability	✓			✓			✓				✓	
• Social sustainability	✓			✓				✓			✓	
• Economic sustainability	✓			✓				✓			✓	
3. Hard smartness	3			1	4			2	6	4		3
• Digitalization of infrastructure	✓				✓				✓		✓	
• Digitalization in tourism	✓				✓				✓		✓	
• Smart technologies for sustainable development	✓				✓			✓			✓	

(*Continued*)

Table C.1.   (*Continued*)

Indicators	Varna			Burgas			Thessaloniki			Dubrovnik		
	A	B	C	A	B	C	A	B	C	A	B	C
4. Soft smartness	1	4		2	2			6			6	
• Creativity		✓		✓				✓			✓	
• Smartness of the community		✓		✓				✓			✓	
• Innovation related to sustainable development	✓				✓			✓			✓	
5. Smart ecosystem	1	4		2	2			6			2	6
• Quality of life		✓		✓				✓				✓
• Quality of tourist experience	✓				✓			✓			✓	
• Attractiveness of the destination		✓		✓				✓				✓

Note: A (basic), B (added value), C (exclusivity).

Bibliography

Agenda 21 for Travel and Tourism Industry (1999). "Towards Environmental Sustainable Development." Commission for Sustainable Development, 7th Session, April 18–May 3, 1999, New York, http://www.un.org/esa/sustdev/sdissues/tourism/tourism_decisions.htm.

Alonso Dos Santos, M., Torres, E., Arboleda, D.Q., Trujilo, E. (2021). "The role of experience and trustworthiness on perception sustainable touristic destinations." *Journal of Hospitality and Tourism Management* **49**: 471–480.

Anastasiadou, K. and Vougias, S. (2019). "'Smart' or 'Sustainably Smart' Urban Road Networks? The Most Important Commercial Street in Thessaloniki as a Case Study." *Transport Policy* **82**: 18–25.

Bastides, A., *et al.* (2020). "The Past, Present, and Future of Smart Tourism Destinations: A Bibliometric Analysis." *Journal of Hospitality and Tourism Research* **45**(3): 520–529.

Bieger, T. (2002). *Management von Destination.* München & Wien: R. Oldenbourg Verlag, Aufgabe 5.

Bieger, T., Müller, H., and Elsasser, H. (2001). "Nachhaltigkeitbegleitung bei der Alpinen Ski-WM 2003." *Tourismus Journal* **1**: 61–75.

Boes, K., Buhalis, D., and Inversini, A. (2015). "Conceptualising Smart Tourism Destination Dimensions." In J. L. Stienmetz, B. Ferrer-Rosell, and D. Massimo (eds.), *Information and Communication Technologies in Tourism*, pp. 391–403. Cham: Springer.

Boes, K., *et al.* (2015). "Smart Tourism Destinations: Ecosystems for Tourism Destination Competitiveness." *International Journal of Tourism Cities* **2**(2): 108–124.

Borruso, G. and Balletto, G. (2022). "The Image of the Smart City: New Challenges." *Urban Science* **6**(1): 5.

Bosselman, F. P., Peterson, C. A., and McCarthy, C. (1999). *Managing Tourism Growth*. Washington, DC: Inland Press.

Buchalis, D. and Fletcher, J. (1998). "Environmental Impacts on Tourist Destinations: An Economic Analysis." In H. Coccossis and P. Nijkamp (eds.), *Sustainable Tourism Development*. Farnham: Ashgate.

Buhalis, D. (2000). "Marketing the Competitive Destination of the Future." *Tourism Management* **21**(1): 97–116.

Buhalis, D. and Amaranggana, A. (2015). "Smart Tourism Destinations: Enhancing Tourism Experience Through Personalisation of Services." In I. Tussyadiah and A. Inversini (eds.), *Information and Communication Technologies in Tourism*, pp. 377–389. Heidelberg, Germany: Springer.

Buhalis, D., Xiang. Z., and Tussyadiah, I. (2015). "Smart Destinations: Foundations, Analytics, and Applications." *Journal of Destination Marketing & Management* **4**(3): 143–144.

Butler, R. (1993). "Sustainable Tourism: Looking Backwards in Order to Progress." In C. M. Hall and A. A. Lew (eds.), *Sustainable Tourism: A Geographical Perspective*. Harlow: Addison Wesley Longman.

Calderón, M., López, G., and Marín, G. (2018). "Smartness and Technical Readiness of Latin American Cities: A Critical Assessment." *IEEE Access* **6**: 8485791.

Carić, H. (2011). "Cruising Tourism Environmental Impacts: Case Study of Dubrovnik, Croatia." *Journal of Coastal Research* **61**: 104–113.

Centre of Regional Science, Vienna UT (2007). *Smart Cities — Ranking European Medium-sized Cities*.

Coca-Stefaniak, J. A. (2019). "Marketing Smart Tourism Cities: A Strategic Dilemma." *International Journal of Tourism Cities* **5**(4): 513–518.

Coccossis, H. and Nijkamp, P. (1995). *Sustainable Tourism Development*. Farnham: Ashgate.

Cooper, C. and Archer, B. (1994). "The Positive and Negative Impacts of Tourism." In William F. Theobald (ed.), *Global Tourism: The Next Decade*. Oxford: Butterworth-Heinemann.

Cooper, C., Fletcher, J., Gilbert, D., Shepherd, S., and Wanhil, R. (1993). *Tourism Principles & Practice*. London: Pitman Publishing.

Dyer, P., Aberdeen, L., and Schuler, S. (2003). "Tourism Impacts on an Australian Indigenous Community: A Djabugay Case Study." *Tourism Management* **24**(1): 83–95.

European Commission (2006). *Methodical Work on Measuring the Sustainable Development of Tourism: Manual on Indicators Sustainable Development of Tourism*. Part 1, Technical Report, Eurostat.

European Commission (2022a). "EU Guide on Data for Tourism Destinations." Smart Tourism Destination, SI2.843962.

European Commission (2022b). "Study on Mastering Data for Tourism by EU Destinations: Methodological Appendix." Publications Office of the European Union, https://op.europa.eu/en/publication-detail/-/publication/9df86541-fba5-11ec-b94a-01aa75ed71a1.

Femenia-Serra, F. and Perea-Medina, M. (2016). "Analysis of Three Spanish Potential Smart Tourism Destinations." In *6th International Conference on Tourism (ICOT)*, Naples.

Gajdošík, T. (2018). "Smart Tourism: Concepts and Insights from Central Europe." *Czech Journal of Tourism* **7**(1): 25–44.

González-Reverte, F. (2019), "Building Sustainable Smart Destinations: An Approach Based on the Development of Spanish Smart Tourism Plans." *Sustainability* **11**(23): 39–43.

Goodall, B. and Stabler, M. J. (1997). "Principles Influencing the Determination of Environmental Standards for Sustainable Tourism." In M. J. Stabler (ed.), *Tourism Sustainability: Principles and Practice*. Washington, DC: CAB International.

Gretzel, U. (2018). "From Smart Destinations to Smart Tourism Regions." *Journal of Regional Research* **42**: 171–184.

Gretzel, U. and Scarpino Johns, M. (2018). "Destination Resilience and Smart Tourism Destinations." *Tourism Review International* **22**(3): 263–276.

Gretzel, U., Zhong, L., and Koo, C. (2016). "Application of Smart Tourism to Cities." *International Journal of Tourism Cities* **2**(2). https://doi.org/10.1108/IJTC-04-2016-0007.

Gretzel, U., Sigala, M., Xiang, Z., and Koo, C. (2015). "Smart Tourism: Foundations and Developments." *Electronic Markets* **25**(3): 179–188.

Hall, D. (2000). "Sustainable Tourism Development and Transformation in Central and Eastern Europe." *Journal of Sustainable Tourism* **8**(6): 441–457.

Holden, A. (2001). *Environment and Tourism*. London: Routledge.

Horn, C. and Simmons, D. (2002). "Community Adaptation to Tourism: Comparisons Between Rotorua and Kaikoura, New Zealand." *Tourism Management* **23**: 133–143.

Hsu, C. C. and Tsaih, R. H. (2018). "Artificial Intelligence in Smart Tourism: A Conceptual Framework." In *Proceedings of the 18th International Conference on Electronic Business (ICEB)*, December 2–6, 2018, Guilin, China, pp. 124–133.

Hunter, C. and Green, H. (1995). *Tourism and the Environment: A Sustainable Relationship?* London: Routledge.

Institute for Market Economics (2020). "Varna — City of Knowledge: Economic Profile and Development Trajectory" (in Bulgarian).

Ivars-Baidal, S. and Giner Sánchez, D. (2016). "Gestión turística y tecnologías de la información y la comunicación (TIC): El nuevo enfoque de los destinos inteligentes." *Documents d'Anàlisi Geogràfica* **62**(2): 327–346.

Ivars-Baidal, J. A., Celdrán-Bernabeu, M. A., Mazón, J. N., and Perles Ivars, A. (2019). "Smart Destinations and the Evolution of ICTs: A New Scenario for Destination Management?" *Current Issues in Tourism* **22**(13): 1581–1600.

Ivars-Baidal, J. A., Celdrán-Bernabeu, M. A., Femenia-Serra, F., Perles-Ribes, J. F., and Giner Sánchez, D. (2021), "Measuring the Progress of Smart Destinations: The Use of Indicators as a Management Tool." *Journal of Destination Marketing & Management* **19**: 100531.

Kercher, B. M. (1993). "The Unrecognized Threat to Tourism: Can Tourism Survive 'Sustainability'." *Tourism Management* **14**(2): 131–136.

Lamsfus, C., Martín, D., Alzua-Sorzabal, A., and Torres-Manzanera, E. (2015). "Smart Tourism Destinations: An Extended Conception of Smart Cities Focusing on Human Mobility." In J. L. Stienmetz, B. Ferrer-Rosell, and D. Massimo (eds.), *Information and Communication Technologies in Tourism*, pp. 363–375. Cham: Springer.

Leiper, B. and Buhalis, D. (1995). "Marketing the Competitive Destination of Future." *Tourism Management* **21**(1).

Li, Y., *et al.* (2017). "The Concept of Smart Tourism in the Context of Tourism Information Services." *Tourism Management* **58**: 293–300.

Lopez de Avila, A. (2015). "Smart Destinations: XXI Century Tourism." In *ENTER2015 Conference on Information and Communication Technologies in Tourism*, LugaNo, Switzerland.

Maráková, V., Dzúriková, L., and Timko, M. (2022). "Developing a Smart Destination: Insights from Slovakia." In *Proceedings of the 5th International Conference on Tourism Research*, pp. 234–244.

Marinov, V. (1998). *The Challenges to the Sustainable Development of Bulgaria.* Yearbook of SU Kliment Ohridski, Geography and Geology Faculty, Volume 2 (in Bulgarian).

Marinov, S. (2003). "Marketing Approach for the Development of Integrative Tourism on the Territory of the Golden Sands Natural Park." In *Varna: Collection of Reports from the Scientific and Practical Conference*, October 30–31, 2003 (in Bulgarian).

Marinov, S., Kazandzhieva, V., *et al.* (2023). *Monitoring the Sustainable Development of a Seaside Tourist Destination.* Newcastle Upon Tyne: Cambridge Scholars Publishing.

McCool, S. F., Moisley, R. N., and Nickerson, P. N. (2009). *Tourism, Recreation and Sustainability: Linking Culture and the Environment.* School of Forestry, University of Montana.

Mihalic, T. and Kaspar, C. (1996). *Umweltökonomie im Tourismus*. St. Gallen: Haupt.

Miller, G. (2001). "The Development of Indicators for Sustainable Tourism: Results of a Delphi Survey of Tourism Researchers." *Tourism Management* **22**: 351–362.

Moura, F. and de Abreu e Silva, J. (2019). "Smart Cities: Definitions, Evolution of the Concept and Examples of Initiatives." In W. Leal Filho, A. Azul, L. Brandli, P. Özuyar, and T. Wall (eds.), *Industry, Innovation and Infrastructure, Series: Encyclopedia of the UN Sustainable Development Goals*, pp. 1–9. Switzerland, AG: SpringerNature.

Murphy, P. (1994). "Tourism and Sustainable Development." In F. William (ed.), *Global Tourism the Next Decade*. Oxford: Heinemann Ltd.

Newsome, D., Moore S. A., and Dowling, R. K. (2002). *Natural Area Tourism: Ecology, Impacts and Management*. Bristol: Channel View Publications.

O'Reilly, A. M. (1986). "Tourism Carrying Capacity: Concept and Issues." *A. M. Tourism Management* **7**(4): 254–258.

Pearce Douglas, G. (1994). *Tourist Development*, 2nd edition. London: Longman Scientific & Technical.

Pearce, D. W. (1993). "Sustainable Development and Developing Country Economies." In R. Kerry Turner (ed.), *Sustainable Environmental Economics and Management: Principles and Practice*. London: John Wiley & Son.

Peng, G., Nunes, M., and Zheng, L. (2017). "Impacts of Low Citizen Awareness and Usage in Smart City Services: The Case of London's Smart Parking System." *Information Systems and e-Business Management* **15**: 845–876.

Pulijic, I., Knezevic, M., and Segota, T. (2019). *Overtourism: Understanding and Managing Urban Tourism Growth beyond Perceptions, Volume 2: Case Studies*, pp. 40–43.

Rafailova, G. (2005). *A Key Determinant for Sustainable Development of a Tourist Destination*. Sofia: Economic Studies, Book 2 (in Bulgarian).

Rafailova, G. (2020). "Development of Varna as Smart Tourist Destination." In Tourism — Beyond Expectations, Proceedings: Jubilee International Scientific Conference, September 25–26, 2020, 100 Years of UNSS and 30 Years of Cat. Economics of Tourism, pp. 421–434. Sofia: Izd. Complex.

Rafailova, G. and Hadzhikolev, A. (2020a). "Key Aspects of Implementing the Smart city Concept in the Tourism Sector." In *Economic Science, Education and Real Economics: Development and Interaction in the Digital Era: Anniversary International Scientific Conference*, Volume 1, pp. 581–591. Varna, Bulgaria: Science and Economics.

Rafailova, G. and Hadzhikolev, A. (2020b). "Assessment of Smart Experience of Tourists and Local Citizens in Tourist Destination." In *Tourism and Connectivity: Proceedings of the Jubilee Scientific Conference with*

International Participation on the Occasion of the 55th Anniversary of the Establishment of the Special Tourism, October 30–31, 2020, pp. 563–569. Varna, Bulgaria: Science and Economics.

Rafailova, G., Todorova-Hamdan, Z., and Filipova, H. (2022). "Economic Sciences Series: Opportunities for Development of Varna as a Smart Tourism Destination (Based on an Expert Survey Among Representatives of the Tourism Industry in Varna)." *Izvestia Journal of the Union of the Scientists — Varna* **11**(1): 10–17.

Rafailova, G., Todorova-Hamdan, Z., and Filipova, H., (2023). "Development of a Human-Centric Model for Assessment of Smart and Sustainable Tourism Destination." *Economic Studies* **32**(2): 151–171.

Rakadzhiyska, S. (1997). "Concept of Sustainable Tourism and the Positions of Bulgaria in the International Markets." *Magazine Izvestiya of UE-Varna,* pp. 33–44 (in Bulgarian).

Rayan, C. (2002). "Equity, Management, Power Sharing and Sustainability — Issues of the 'New Tourism.'" *Tourism Management* **23**: 17–26.

Ringel, M. (2021). "Smart City Design Differences: Insights from Decision-Makers in Germany and the Middle East/North-Africa Region." *Sustainability* **13**(4): 2143.

Segittur (2015). *Smart Tourist Destinations Report: Building the Future.* Madrid: Smart Destination.

Shafiee, S., *et al.* (2019). "Developing a Model for Sustainable Smart Tourism Destinations: A Systematic Review." *Tourism Management Perspectives* **31**: 287–300.

Simpson, K. (2001). "Strategic Planning and Community Involvement as Contributors to Sustainable Tourism Development." *Current Issues in Tourism* **4**(1): 3–41.

TNO Inro — Department of Spatial Development in Co-operation with CISET Venezia, Italy, and the University of Innsbruck, Austria (2002). "Early Warning System for Identifying Declining Tourist Destinations and Preventive Best Practice." Luxembourg: Office for Official Publications of the European Communities.

Todorova-Hamdan, Z. and Hadzhikolev, A. (2021). *Application of Information and Communication Technologies in Tourism: Collection of Reports from the Round Table on the Topic Modern Tourism. Smart Solutions for the Development of Tourism in Bulgaria in the Conditions of the COVID-19 Pandemic.* Science and Economics by UE-Varna (in Bulgarian).

Urry, J. (1994). *The Tourist Gaze: Leisure and Travel Contemporary Society.* London: Sage.

Valentine, P. S. (1993). "Ecotourism and Nature Conservation: A Definition with Some Recent Developments in Micronesia." *Tourism Management* **2**.

Vargas-Sánchez, A. (2016). "Exploring the Concept of Smart Tourist Destination." *Enlightening Tourism: A Pathmaking Journal* **6**(2): 178–196.

Vodenska, M. (2001). *Economic, Social and Natural Impacts of Tourism.* Sofia: Kliment Ohridski University Publishing House (in Bulgarian).

World Tourism Organization and International Transport Forum (2019). *Transport-related CO_2 Emissions of the Tourism Sector — Modelling Results.* Madrid: UNWTO.

Yigitcanlar, T., Kamruzzaman, M., Foth, M., Sabatini-Marques, J., da Costa, J., and Ioppolo, G. (2019). "Can Cities Become Smart Without Being Sustainable? A Systematic Review of the Literature." *Sustainable Cities and Society* **45**: 348–365.

Internet Sources

Allo Mairie Nice (2023). www.nice.fr/fr/allo-mairie (Accessed 12 February 2023).

Batchelder, C. (2018). Stunning and SMART: Why Dubrovnik Is Turning Heads about Digital, TechPlace, https://www.techplace.online/stunning-and-smart-why-dubrovnik-is-turning-heads-about-digital/ (Accessed 14 March 2023).

Black Sea CBC (2021). Promoting Heritage and Culture-based Experiential Tourism in the Black Sea Basin, 62–66, https://blacksea-cbc.net/wp-content/uploads/2021/06/BSB1145_PRO-EXTOUR_-_Regional-needs-assessment-report-for-the-development-of-experiential-tourism-in-the-involved-countries_EN.pdf (Accessed 10 February 2023).

Capital (2023). Passengers at the Airports in Varna and Burgas are 59% More than Last Year, https://www.capital.bg/biznes/transport/2023/01/08/4435615_putnicite_na_letishtata_vuv_varna_i_burgas/ (Accessed 13 March 2023).

Chatbot (2023). Ville de Nice, https://chatbot.nice.fr/ (Accessed 10 March 2023).

Chrysostomou, K. (2015). Smart and Resilient Urban Mobility Planning in Thessaloniki, https://civitas.eu/sites/default/files/wiki_pre_sgv_hit_resilience.pdf (Accessed 20 March 2023).

City of Thessaloniki (2023). https://thessaloniki.gr/?lang=en (Accessed 20 March 2023).

Croatia Week (2013). Dubrovnik: A City For All Seasons, Presented In London, https://www.croatiaweek.com/dubrovnik-a-city-for-all-seasons-presented-in-london/ (Accessed 20 February 2023).

Dubrovačko Oko (2023). Osnovna Statistika, http://dubrovacko-oko.hr/statistics (Accessed 12 February 2023).

Dubrovnik Airport (2023). Statistics, https://www.airport-dubrovnik.hr/en/business/statistics-s36 (Accessed 13 March 2023).

Earth Observation Toolkit for Sustainable Cities and Human Settlements (2023). The Urban Resilience Observatory for the Evaluation of SDGs of the Municipality of Thessaloniki, https://eo-toolkit-guo-un-habitat.opendata.arcgis.com/pages/thessaloniki-use-case (Accessed 11 February 2023).

European Commission (2016). The European Tourism Indicator System ETIS Toolkit for Sustainable Destination Management, http://ec.europa.eu/growth/sectors/tourism/offer/sustainable/indicators/index_en.htm (Accessed 20 January 2023).

European Commission (2021). Cultural and Creative Cities Monitor, https://composite-indicators.jrc.ec.europa.eu/cultural-creative-cities-monitor/performance-map (Accessed 4 March 2023).

European Commission's Intelligent City Challenge (2023). Nice, France, https://www.intelligentcitieschallenge.eu/cities/nice (Accessed 15 January 2023).

Eurostat (2021). Seasonality in Tourism Demand, https://ec.europa.eu/eurostat/statistics-explained/index.php?title=Seasonality_in_tourism_demand (Accessed 20 June 2022).

Future Market Magazine (2023). Spanish Smart City: Valencia, https://future-markets-magazine.com/en/markets-technology-en/smart-city-valencia/ (Accessed 16 March 2023).

GeoLand (2023). Geocultura is the 8th Biennale of Contemporary Art, https://www.geolandproject.eu/2023/01/15/geocultura/ (Accessed 12 February 2023).

Geospatial Enabling Technologies (2023). Application of Performance Indicators for Sustainable Smart Cities of the ISO 37120 Series — Municipality of Thessaloniki, https://www.getmap.eu/project/thessaloniki-kpi/?lang=en (Accessed 11 February 2023).

Greek Travel Pages (2022). Thessaloniki Sets Goal for Year-round Tourism, https://news.gtp.gr/2022/08/23/thessaloniki-sets-goal-for-year-round-tourism/ (Accessed 12 March 2023).

IBM Smarter Cities Challenge (2011). Nice, France, https://www.smartercitieschallenge.org/city_nice_france.html (Accessed 12 February 2023).

IBM Smarter Cities Challenge (2015). Thessaloniki, Greece, https://www.smartercitieschallenge.org/cities/thessaloniki-greece/ (Accessed 12 February 2023).

iNews (2022). Summer 2022 Remains Without a Single Cruise Ship in the Ports of Burgas and Varna, https://inews.bg/Туризъм/Лято-2022-остава-без-нито-един-круизен-кораб-на-пристанищата-в-Бургас-и-Варна_l.a_c.362_i.695543.html (Accessed 10 March 2023).

Intelligent Cities Challenge (2019). Digital Transformation Strategy for the City of Thessaloniki, https://www.intelligentcitieschallenge.eu/sites/default/

files/2019-07/Digital_transformation_strategy_THESSALONIKI.pdf (Accessed 11 February 2023).

Investment Portal of Burgas Municipality (2023a). *Macroeconomic Profile*, https://invest.burgas.bg/bg/makroikonomicheski-profil (Accessed 10 January 2023).

Investment Portal of Burgas Municipality (2023b). *Culture and Tourism*, https://invest.burgas.bg/bg/kultura-i-turizam (Accessed 10 January 2023).

Investment Portal of Burgas Municipality (2023c). *Quality of Life*, https://invest.burgas.bg/bg/kachestvo-na-zhivot (Accessed 10 January 2023).

L'Institut national de la statistique et des études économiques (2023). Statistiques et études, www.insee.fr/fr/statistiques (Accessed 10 February 2023).

Lignes d'Azur (2023). Public Transport in Nice, www.lignesdazur.com/lignes-dazur-sengage-pour-vous-avec-ses-reseaulutions (Accessed 10 February 2023).

Major Development Agency Thessaloniki (MDAT) (2021). Action Plan for Thessaloniki, https://projects2014-2020.interregeurope.eu/fileadmin/user_upload/tx_tevprojects/library/file_1620813428.pdf (Accessed 14 March 2023).

Métropole Nice Côte d'Azur (2023). www.nicecotedazur.org (Accessed 11 February 2023).

Municipality of Burgas (2018). Burgas Hosts a Forum on the Mobility of EUROCITIES, https://www.burgas.bg/bg/posts/view/35866?pdf=1 (Accessed 14 March 2023).

Municipality of Burgas (2019). Today is the Official Launch of the Integrated Urban Platform Smart Burgas, https://www.burgas.bg/bg/posts/view/40520/ (Accessed 15 March 2023).

Municipality of Varna (2022a). Integrated Development Plan of Varna Municipality 2021–2027, https://www.varna.bg/bg/1612 (Accessed 29 March 2023).

Municipality of Varna (2022b). Program for the Development of Tourism in the Municipality of Varna 2021–2027 г., https://www.varna.bg/bg/1663 (Accessed 22 March 2023).

Municipality of Varna (2022c). Programs, https://www.varna.bg/343 (Accessed 10 March 2023).

Municipality of Varna, Directorate of "Tourism" (2022). Tourism Analysis of Destination Varna, https://www.varna.bg/bg/164 (Accessed 9 March 2023).

Municipality of Varna, Smart and Sustainable Cities (mySMARTLife) (2019). https://www.varna.bg/bg/380 (Accessed 14 March 2023).

NSI (2021). National Statistical Institute of Bulgaria, https://www.nsi.bg/bg (Accessed 30 March 2023).

NSI (2022). "Population as of 31.12.2021 by Regions, Municipalities, Place of Residence and Sex." https://www.nsi.bg/bg/content/2975/население-по-области-общини-местоживеене-и-пол (Accessed 12 March 2023).

NSI (2023). GDP by Region, https://www.nsi.bg/bg/content/2215/бвп-регионално-ниво (Accessed 17 March 2023).

Numbeo (2023), Quality of Life, https://www.numbeo.com/quality-of-life/ (Accessed 15 March 2023).

Office du Tourisme Métropolitan Nice Côte d'Azur. (2023) Официален сайт на ТИЦ Ница, www.explorenicecotedazur.com (Accessed 10 February 2023).

PIRO Burgas 2021–2027 (2022). Plan for the Integrated Development of the City of Burgas (2021–2027), http://plan.smartburgas.eu/пиро-2021-2027 (Accessed 10 February 2023).

Plastic Smart Cities (2021). Action Plan to Reduce Plastic Pollution in the City of Dubrovnik 2021–2026, https://www.dubrovnik.hr/uploads/posts/14644/Action-Plan-to-reduce-the-plastic-pollution-in-the-City-of-Dubrovnik-2021.-2026..pdf (Accessed 15 February 2023).

Provence-Alpes-Côte d'Azur (2023). Top des visites guides, https://provence-alpes-cotedazur.com/que-faire/visites-guidees/top-des-visites-guidees-incontournables-a-faire-a-nice/ (Accessed 11 February 2023).

RIOSV (2014). List of Protected Areas, https://riosv-varna.bg/zashtiteni-teritorii-i-bioraznoobrazie/zashtiteni-teritorii/spisak-na-zashtiteni-teritorii-vav-varnenska-oblast/ (Accessed 22 March 2023).

RIOSV (2021a). Regional Report on the State of the Environment 2021, https://www.riosv-varna.bg/docs/RIOSV-Varna%20Doklad%20OS-2021.pdf (Accessed 10 March 2023).

RIOSV (2021b). State of the Environment, https://riosv-varna.bg/info-centar/sastoyanie-na-okolnata-sreda/ (Accessed 15 March 2023).

Rogulj, D. (2022). 520,000 Cruise Ship Passengers in Dubrovnik Expected this Season, *Total Croatia News*, https://total-croatia-news.com/news/travel/cruise-ship-passengers-in-dubrovnik/ (Accessed 12 March 2023).

Sharing Cities (2022). To Create Smart Cities Together, https://sharingcities.eu/wp-content/uploads/sites/6/2022/07/SharingCitiesLeaflet_BG_final_web.pdf (Accessed 16 March 2023).

SKG Airport Greece (2023). Thessaloniki airport 'Makedonia' — 2022 vs 2021, https://www.skg-airport.gr/uploads/sys_nodelng/2/2869/Thessaloniki_12_Traffic_2022vs2021.pdf (Accessed 13 March 2023).

SmartBurgas (2023). Integrated Urban Platform Burgas, https://smartburgas.eu/ (Accessed 11 January 2023).

Spain National Statistics Institute (2021). https://www.ine.es/en/ (Accessed 20 February 2023).

Statista (2022). Information Technology (IT) Spending Forecast Worldwide from 2012 to 2023, by Segment, https://www.statista.com/statistics/268938/global-it-spending-by-segment/ (Accessed 24 August 2022).

Stieghorst, T. (2017). Dubrovnik Seeks to Sharply Curtail Cruise Tourism, *Travel Weekly*, https://www.travelweekly.com/Cruise-Travel/Dubrovnik-seeks-to-sharply-curtail-cruise-tourism (Accessed 18 February 2023).

Suspanish (2023). Valenbisi Valencia, https://www.suspanish.com/blog/valenbisi-valencia/ (Accessed 15 March 2023).

Thessaloniki Airport (2023). Statistics for Thessaloniki Airport, https://thessalonikiairportmakedonia.com/statistics/ (Accessed 13 March 2023).

UNESCO World Heritage Convention (2021). Nice, Winter Resort Town of the Riviera, https://whc.unesco.org/en/list/1635/ (Accessed 12 February 2023).

ValenciaPort (2023). Terminals, www.valenciaport.com/en/passengers/cruises/terminals/ (Accessed 11 January 2023).

Ville de Nice (2023). www.nice.fr (Accessed 11 February 2023).

Visit Valencia (2022). *Valencia, a Smart Destination*, https://www.visitvalencia.com/en/smart-tourism (Accessed 17 March 2023).

WeGov (2017). *Smart City Dubrovnik*, http://we-gov.org/wp-content/uploads/2017/11/9-ANA-MARIA-Smart-City-Dubrovnik.pdf?ckattempt=1 (Accessed 10 March 2023).

World Population Review (2022). *Thessaloniki Population 2023*, https://worldpopulationreview.com/world-cities/thessaloniki-population (Accessed 10 March 2023).

Index

B
blue economy, 152

C
citizens' opinions, 62
collective mobile applications, 67
community intelligence, 55
comparative characteristics, 127
creative industries, 54
creative tourist destination, 152
crowd control, 72

D
digitalization, 20

E
economic sustainability, 39
e-government, 53
environmental impact management, 54
environmental sustainability, 21
e-tourism, 16

G
green innovations, 54

H
hard smartness, 15
"hardware" smartness, 48
human-centric approach, 46

I
innovative and sustainable maritime tourism, 152
intermodal transport, 68

M
methodology for assessing a smart and sustainable tourist destination, 51

N
negative and positive influences of tourism, 31

O
open database, 53
opinion of experts, 62

P
public–private partnership, 71

Q
quality of life, 8

S
smart and sustainable destination, 46
smart business, 22
smart city, 1
smart ecosystem, 48
"smart" experience, 11
smart governance, 20
smart mobility, 22
smart supply, 95
smart systems, 20
smart technologies, 55
smart tourism, 14
smart tourism destination, 1

social entrepreneurship, 54
social integration, 54
social sustainability, 39
soft smartness, 16
"software" smartness, 48
stakeholders, 107
sustainable destination development, 22

T
theory of the S-D logic, 17
tourists' opinion, 62

V
valuable experience for tourists, 55
value co-creation, 16